U.S. Department of Transportation

BUDGET ESTIMATES

FISCAL YEAR 2012

RESEARCH AND INNOVATIVE TECHNOLOGY ADMINISTRATION

SUBMITTED FOR THE USE OF
THE COMMITTEES ON APPROPRIATIONS

Research and Innovative Technology Administration
FY 2012 Congressional Budget Request
Table of Contents

SECTION 1: OVERVIEW

ADMINISTRATOR'S OVERVIEW	1
Exhibit I: ORGANIZATION CHART	5

SECTION 2: BUDGET SUMMARY TABLES

Exhibit II-1:	Comparative Statement of New Budget Authority	7
Exhibit II-2:	Budget Request by Appropriation Account	8
Exhibit II-3:	Budget Request by DOT Strategic and Organizational Goals	10
Exhibit II-3A:	Budget Request by DOT Outcomes	11
Exhibit II-4:	Budgetary Authority	14
Exhibit II-5:	Outlays	15
Exhibit II-6:	Summary of Requested Funding Changes from Base	16
Exhibit II-7:	Working Capital Fund	18
Exhibit II-8:	Personnel Resource Summary (FTE)	19
Exhibit II-9:	Personnel Resource Summary (FTP)	20

SECTION 3: BUDGET REQUEST BY APPROPRIATION ACCOUNT

Research and Development:

Appropriation Language	21
Exhibit III-1: Appropriation Summary by Program Activity	22
Exhibit III-1a: Summary Analysis of Change	23
Exhibit III-2: Annual Performance Results and Targets	24
Program and Performance	27
Program and Financing	28
Object Classification	29
Funding History	30

Detailed Justifications by Program Activity:

Salaries and Administrative Expenses	31
Alternative Fuels Research and Development (R&D)	32
RD&T Coordination	36
Nationwide Differential Global Positioning System	41
Positioning, Navigation and Timing	45

Reimbursable Programs/Other:

Competitive University Transportation Centers (UTC) Program	51
Competitive University Transportation Center (UTC) Consortia	51
University Transportation Center (UTC) Multimodal Competitive Research Grants	52
Multimodal Innovative Research Program	53
Transportation Safety Institute	54
Volpe National Transportation Systems Center	55
Intelligent Transportation Systems	59

Research and Innovative Technology Administration
FY 2012 Congressional Budget Request
Table of Contents

Bureau of Transportation Statistics:

Exhibit III-1: Appropriation Summary by Program Activity	63
Summary by Program and Administrative Expenses	64
Exhibit III-1a: Summary Analysis of Change	65
Exhibit III-2: Annual Performance Results and Targets	66
Object Classification	68
Funding History	69
Detailed Justifications by Program Activity:	
Safety Data and Analysis Program	70
Passenger Travel Statistics	74
Freight Statistics	79
Transportation Economics	85
Geospatial Information Systems	89
Transportation Analysis, Data Quality and Performance Measurement	92
National Transportation Library	95
Reimbursable Program:	
Airline Transportation Statistics	99
SECTION 4: RESEARCH, DEVELOPMENT AND TECHNOLOGY	
Exhibit IV-1: Budget Authority	105

RESEARCH AND INNOVATIVE TECHNOLOGY ADMINISTRATION (RITA)
FY 2012 Congressional Budget Submission

ADMINISTRATOR'S OVERVIEW

RITA's FY 2012 budget request is $52.6 million, an increase of $12.6 million over the FY 2010 level. This request reflects RITA's Surface Transportation Reauthorization proposal to greatly enhance the data collection and statistical analysis program in support of data-driven decision-making across the USDOT. It will enable RITA to:

- Provide robust transportation statistics research, analysis, and reporting in support of USDOT and stakeholder needs.
- Coordinate, facilitate, review and maximize the effectiveness of the USDOT's research, development and technology (RD&T) portfolio.
- Advance innovative technologies in an increasingly multi-modal transportation system and further education and training in transportation and related fields.

As the USDOT's research and innovation focal point, RITA advances DOT strategic goals by working across the modal administrations and collaborating with partners from other federal agencies, state and local governments, universities, stakeholder organizations, transportation professionals and system operators. From conducting advanced research to providing funding to test and evaluate new approaches across modes; from collecting and analyzing data to training transportation professionals, RITA enables and accelerates transportation innovation.

HIGHLIGHTS OF THE REQUEST

Quality Data & Analysis for Sound Decision-Making

The Bureau of Transportation Statistics (BTS) develops and disseminates timely, relevant and high quality data and information for all transportation modes. Users include federal, state and local agencies, researchers, and public and private sector transportation decision-makers. BTS provides data through programs that address:

- Freight and Travel Statistics (National and International);
- Transportation Economics;
- Geospatial Information Systems;
- Statistical Methods and Standards Performance Metrics; and
- Airline Statistics (Funded through the Research and Development appropriation).

The budget request for BTS is $35 million, an increase of $8 million over the FY 2010 funding level. This represents a new baseline to reflect RITA's Surface Transportation Reauthorization proposal to substantially enhance the data collection and rigorous statistical analysis programs in support of data-driven decision-making that will affect major DOT strategic objectives moving forward, such as advancing transportations safety, driving economic competitiveness through programs such as high speed rail, and ensuring environmental sustainability. This funding will allow RITA to meet DOT's objective of making transportation investments and policy decisions based on sound analytical data.

The $8 million increase to BTS will initiate or provide enhancements to the following programs:

- A new Safety Data and Analysis Program (+$2 million) will support the modal administrations in the areas of safety data collection, statistical expertise, and methodology to improve safety data access, as well as address emerging issues in transportation safety – such as the objective to achieve improvements and consistency in the safety of transit systems across the U.S. Safety data from across the modes will be integrated into a centralized source for cross-modal decision-making, in support of the USDOT Safety Council.

- A bolstered BTS Freight Statistics Program (+$5.9 million) will fund expansion of the highly valued Commodity Flow Survey (CFS), Vehicle Inventory and Use Survey (VIUS) and International Freight Data System (IFDS). These programs collect, compile and analyze freight data across all modes of transportation and provide analytic reports and stakeholder-focused products from multimodal and intermodal perspectives. The CFS data collection will be completed, a methodology for VIUS will be developed, for-hire truck data and targeted surveys will be developed and IFDS data will be delivered to modal customers. These enhanced data collection efforts will provide further refinement to the analysis of freight transportation trends in support of USDOT priorities – particularly enhancing economic competitiveness by driving supply chain efficiency.

Robust Transportation Research

- RITA requests $17.6 million for its Research and Development account, which is a funding increase of $4.6 million over the FY 2010 level. RITA manages and conducts the strategic planning, coordination, facilitation and review of the USDOT's research programs.

- USDOT is responsible for coordinating and developing Positioning, Navigation and Timing (PNT) (+$.600 million) technology, as well as PNT policy coordination and spectrum management. The PNT program enables RITA to fulfill the USDOT's civil PNT leadership role in ensuring federal civil agencies have significant participation in and are adequately represented in the joint management of global positioning systems (GPS). RITA also funds the Nationwide Differential Global Positioning System (NDGPS) program (+$3 million), which provides national, real-time accurate dynamic positioning and navigation information at one-to-three meters for surface transportation users.

Additionally, key components such as the Research Development &Technology (RD&T) (+$.900 million) Planning Council and Planning Team provide critical coordination and leadership among USDOT modal administrations and stakeholders.

These are critical elements for ensuring lab-to-marketplace application of new ideas and technologies.

Critical Components of RITA's Programs

- The University Transportation Center (UTC) program ($100 million) advances U.S. technology and expertise in many transportation-related disciplines, and advances DOT RD&T priorities through baseline funding for university-based transportation education, research and technology transfer as well as a new competitive cross-modal component. The program is jointly funded by reimbursable agreements with the Federal Highway Administration (FHWA) [$72 million] and the Federal Transit Administration (FTA)[$8 million]. An additional $20 million competitive pool funded through the FHWA will be reserved for a targeted collaborative multimodal research program for high priority needs and topics.

- A new Multimodal Innovative Research Program, funded at $20 million, will build on an existing advanced research program and provide cutting-edge research focusing on USDOT priorities under the governance of the RD&T Planning Council. The multi-modal structure of the Planning Council will ensure collaboration across the USDOT. This program is funded by a reimbursable agreement with FHWA.

Cutting-Edge Technologies

The Intelligent Transportation Systems (ITS) program requests $110 million for the ITS core program in FY 2012, along with an additional $100 million increase in conjunction with the President's Wireless Innovation and Infrastructure Initiative, Wireless Innovation (WIN) Fund. This additional $100 million from the WIN fund will be used over a five year period, and will provide the Intelligent Transportation Systems (ITS) program the ability to seek new and innovative opportunities to pursue ground-breaking research and competitive deployments of wireless technology applications. Deployment of the resulting proven ITS applications will greatly advance the safety, efficiency, convenience, and environmental sustainability of surface transportation.

RITA coordinates, facilitates, and reviews well over $1 billion in transportation-related research, analysis, technology transfer, deployment, education and training activities spanning Departmental priorities. The John A. Volpe National Transportation Systems Center and the Transportation Safety Institute, both fee-for-service organizations, provide critical support to meet this mission.

RITA's FY 2012 budget supports all of the Department's strategic goals. It ensures that decision-makers will have access to robust data collection and analysis, advanced research, and cutting-edge technologies.

EXHIBIT I
RESEARCH AND INNOVATIVE TECHNOLOGY ADMINISTRATION
FY 2011

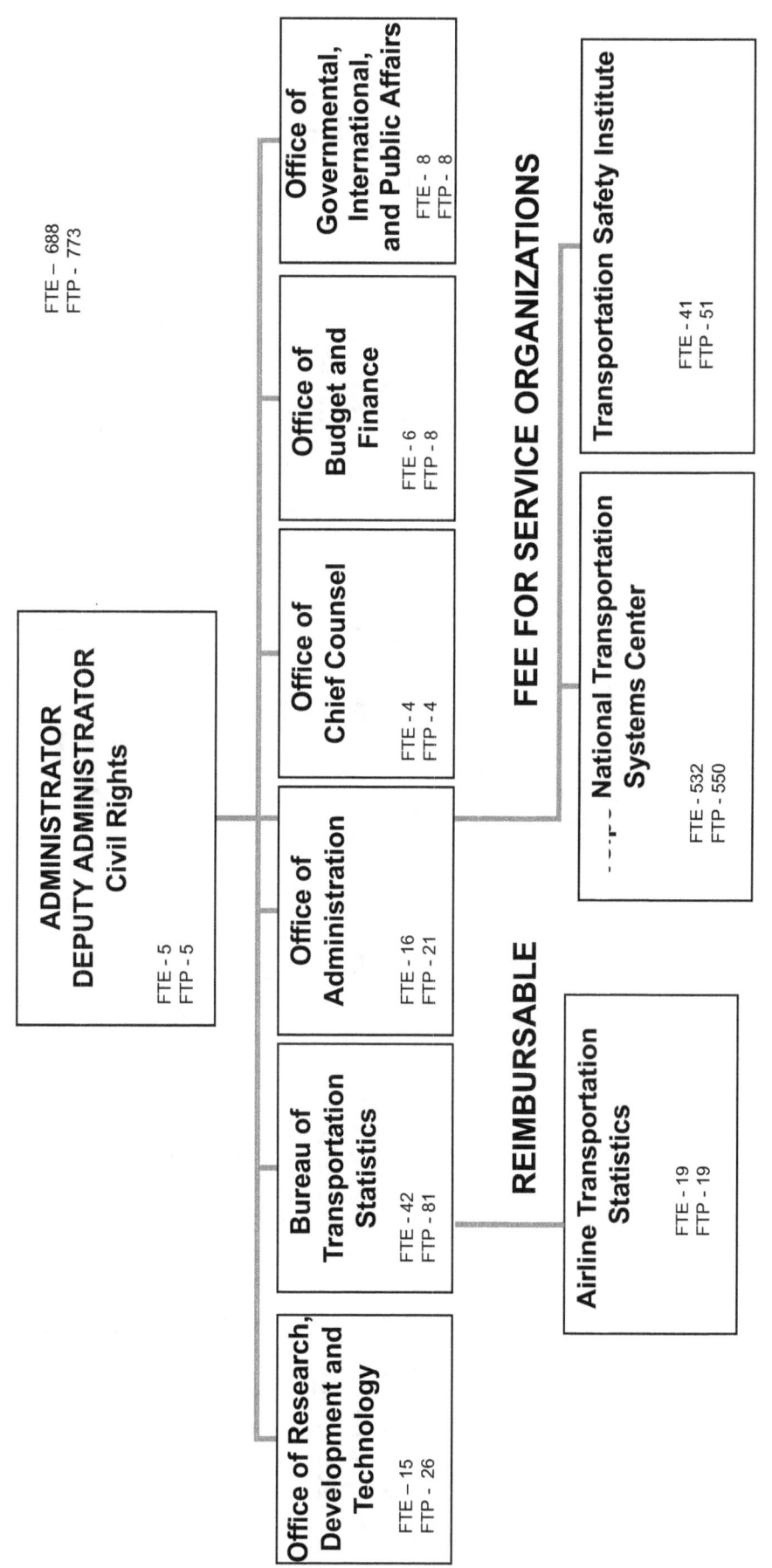

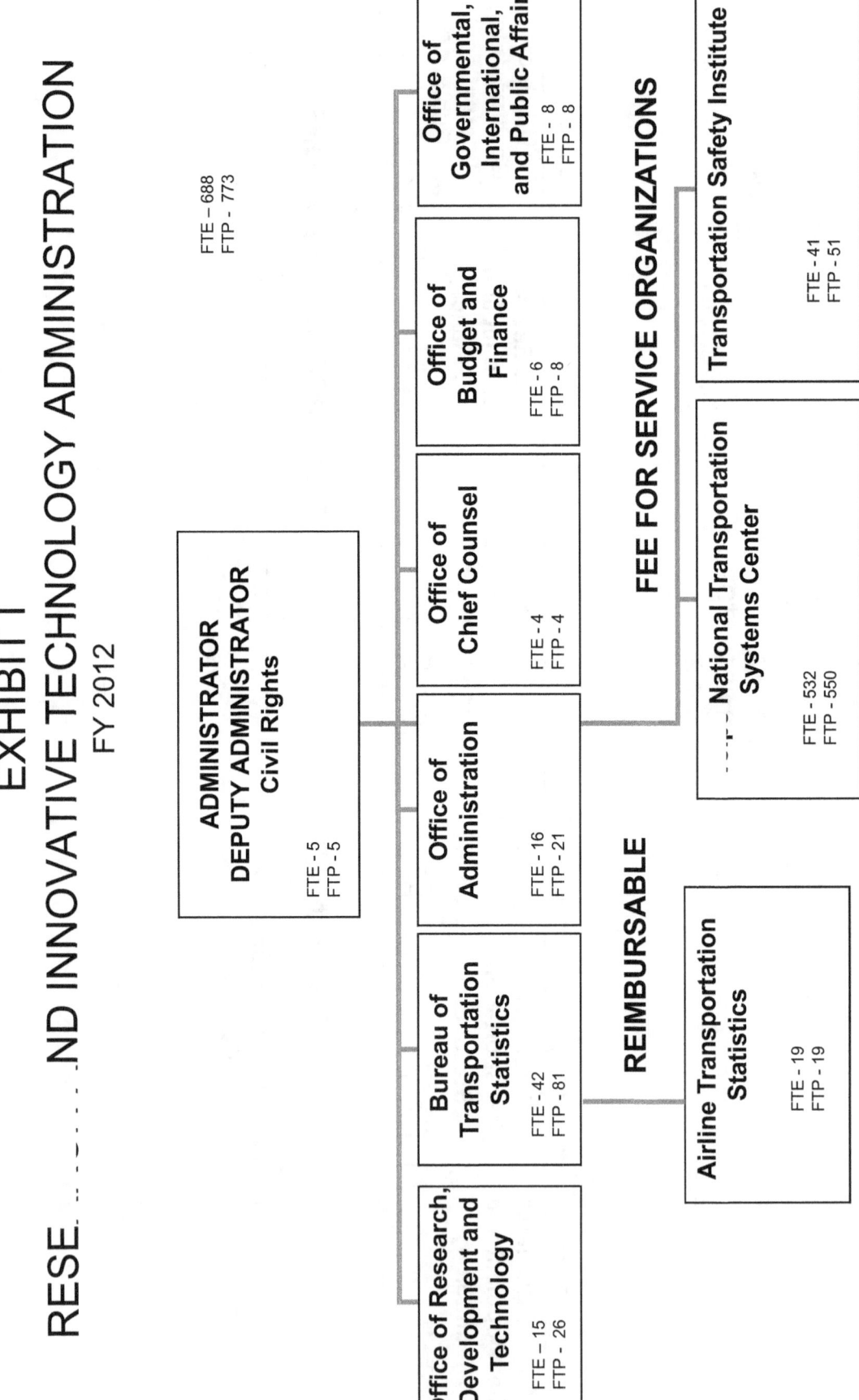

EXHIBIT II-1

FY 2012 NEW BUDGET AUTHORITY
RESEARCH AND INNOVATIVE TECHNOLOGY ADMINISTRATION
($000)

Account Name	FY 2010 ACTUAL	FY 2011 CR (ANNUALIZED)	FY 2012 REQUEST
Research and Development	13,007	13,007	17,600
Bureau of Transportation Statistics (HTF) [1/]	[27,000]	[27,000]	[35,000]
TOTAL	**13,007**	**13,007**	**17,600**

[1/] Resources are shown as non-adds because the Bureau of Transportation Statistics is an allocation account under the Federal-aid Highways program.

EXHIBIT II-2

**FY 2012 BUDGET RESOURCES BY APPROPRIATION ACCOUNT
RESEARCH AND INNOVATIVE TECHNOLOGY ADMINISTRATION
Research and Development
Appropriations, Obligation Limitations and Exempt Obligations
($000)**

ACCOUNT NAME	FY 2010 ACTUAL	FY 2011 CR (ANNUALIZED)	FY 2012 REQUEST
Research and Development:			
Salaries and Administrative Expenses	6,971	6,971	7,600
Alternative Fuels Research & Development (R&D)	500	500	500
RD&T Coordination	536	536	900
Nationwide Differential Global Positioning System	4,600	4,600	7,600
Positioning, Navigation and Timing (PNT)	400	400	1,000
TOTAL: [Discretionary]	**13,007**	**13,007**	**17,600**

EXHIBIT II-2

FY 2012 BUDGET RESOURCES BY APPROPRIATION ACCOUNT
RESEARCH AND INNOVATIVE TECHNOLOGY ADMINISTRATION
Allocation/Reimbursable/Other Programs
Appropriations, Obligation Limitations and Exempt Obligations
($000)

ACCOUNT NAME	FY 2010 ACTUAL	FY 2011 CR (ANNUALIZED)	FY 2012 REQUEST
Allocation/Reimbursable/Other Programs:			
Competitive University Transportation Center (UTC) Consortia (FHWA)	[73,772]	[73,772]	[72,000]
Competitive University Transportation Center (UTC) Consortia (FTA)	[7,000]	[7,000]	[8,000]
University Transportation Center (UTC) Multimodal Competitive Research Grants	[0]	[0]	[20,000]
Multimodal Innovative Research Program	[0]	[0]	[20,000]
Transportation Safety Institute	[20,000]	[20,000]	[20,000]
Volpe National Transportation Systems Center	[250,000]	[250,000]	[250,000]
Intelligent Transportation Systems: Core Program	[102,850]	[102,850]	[110,000]
Intelligent Transportation Systems: Wireless Innovation and Infrastructure Initiative	[0]	[0]	[100,000]
TOTAL: [Allocation/Reimbursable/Other]	[453,622]	[453,622]	[600,000]

EXHIBIT II-2

FY 2012 BUDGET RESOURCES BY APPROPRIATION ACCOUNT
RESEARCH AND INNOVATIVE TECHNOLOGY ADMINISTRATION
Bureau of Transportation Statistics
Appropriations, Obligations Limitations and Exempt Obligations
($000)

ACCOUNT NAME	FY 2010 ACTUAL	FY 2011 CR (ANNUALIZED)	FY 2012 REQUEST
Bureau of Transportation Statistics			
Safety Data and Analysis	0	0	3,049
Travel Statistics	2,947	2,947	3,064
Freight Statistics	10,723	10,723	16,021
Transportation Economics	1,811	1,811	1,938
Geospatial Information Systems	1,758	1,758	1,521
Transportation Analysis, Data Quality and Performance Metrics	7,416	7,416	7,268
National Transportation Library	2,345	2,345	2,139
TOTAL:[1/]	[27,000]	[27,000]	[35,000]
Reimbursable Program:			
Airline Transportation Statistics Program	[4,000]	[4,000]	[5,000]

[1/] Resources are shown as non-adds because the Bureau of Transportation Statistics is an allocation account under the Federal-Aid Highway program.

Exhibit II-3
FY 2012 BUDGET REQUEST BY DOT STRATEGIC AND ORGANIZATIONAL GOALS
RESEARCH AND INNOVATIVE TECHNOLOGY ADMINISTRATION
($000)

	Safety	Environmental Sustainability	State of Good Repair/ Infrastructure	Livable Communities	Economic Competitiveness	Organizational Excellence	Total
Research and Development[1]							
Alternative Fuels Research and Development (R&D)	-	1,376	-	-	-	-	1,376
RD&T Coordination	928	526	495	433	401	3,673	6,456
Research Coordination	618	309	372	278	278	2,449	4,304
Knowledge Management System	186	155	93	93	93	816	1,436
Education & Outreach	62	31	15	31	15	204	358
Technology Transfer	62	31	15	31	15	204	358
Nationwide Differential Global Positioning System (NDGPS)	8,184	-	-	-	-	-	8,184
Positioning, Navigation amd Timing (PNT)	1,267	317	-	-	-	-	1,584
Competitive University Transportation Center (UTC) Consortia	[15,360]	[15,360]	[15,360]	[15,360]	[15,360]	[3,200]	[80,000]
University Transportation Center (UTC) Multimodal Competitive Research Grants	[3,840]	[3,840]	[3,840]	[3,840]	[3,840]	[800]	[20,000]
Multimodal Innovative Research Program	[3,840]	[3,840]	[3,840]	[3,840]	[3,840]	[800]	[20,000]
Intelligent Transportation Systems: Core Program	[52,673]	[6,622]	[0]	[7,204]	[24,101]	[19,400]	[110,000]
Research	[52,673]	[6,622]	-	[7,204]	[24,101]	-	[90,600]
Applications	[39,733]	[6,622]	-	-	[19,866]	-	[66,221]
Technology	[8,940]	-	-	-	[2,235]	-	[11,175]
Policy	[4,000]	-	-	-	-	-	[4,000]
Mode Specific	-	-	-	[4,500]	-	-	[4,500]
Exploratory	-	-	-	-	[2,000]	-	[2,000]
Other/Legacy Initiatives	-	-	-	[2,704]	-	-	[2,704]
Technology Transfer & Evaluation	-	-	-	-	-	[13,900]	[13,900]
Program Support	-	-	-	-	-	[5,500]	[5,500]
Intelligent Transportation Systems: Wireless Innovation and Infrastructure Initiative	[25,000]	[25,000]	[0]	[25,000]	[25,000]	[0]	[100,000]
Bureau of Transportation Statistics[2]	[9,825]	[3,546]	[5,398]	[3,633]	[11,443]	[1,155]	[35,000]
Safety Data and Analysis	[3,049]	-	-	-	-	-	[3,049]
Travel Statistics	[396]	[781]	[196]	[781]	[781]	[129]	[3,064]
Freight Statistics	[4,430]	[1,265]	[3,164]	[633]	[6,326]	[203]	[16,021]
Transportation Economics	-	-	-	-	[1,938]	-	[1,938]
Geospatial Information Systems	[250]	[250]	[250]	[521]	[250]	-	[1,521]
Transportation Analysis, Data Quality and Performance Metrics	[1,348]	[898]	[1,437]	[1,347]	[1,797]	[441]	[7,268]
National Transportation Library	[352]	[352]	[351]	[351]	[351]	[382]	[2,139]
Airline Transportation Statistics[3]	[400]	[400]	[400]	[400]	[3,000]	[400]	[5,000]
TOTAL	10,379	2,219	495	433	401	3,673	17,600

[1] RD&T direct FTE are distributed based on actual FTE allocated to the program. Overhead FTE are distributed to Organizational Excellence.

[2] BTS FTE are distributed proportionately among each goal.

[3] Airline Transportation Statistics FTE is distributed proportionately among each goal.

	Exhibit II-3a		
	FY 2012 BUDGET REQUEST BY DOT OUTCOMES		
	RESEARCH AND INNOVATIVE TECHNOLOGY ADMINISTRATION		
	($000)		

DOT Outcome	Program	FY 2012 Request
SAFETY		
Reduction in injuries and fatalities	BTS Program - Safety Data Analysis (primary)	[1,032]
	BTS Program - Travel statistics (secondary)	[794]
	BTS Program - Freight Statistics (secondary)	[5,555]
	BTS Program - Geospatial Information Systems (secondary)	[238]
	BTS Program - Transportation Analysis, Data Quality, and Performance Metrics (secondary)	[1,190]
	RD&T Program - Research Coordination (primary)	161
	RD&T Program - Knowledge Management System (primary)	121
	RD&T Program - Education & Outreach (primary)	121
	RD&T Program - Technology Transfer (primary)	81
	Nationwide Differential Global Positioning System (primary)	8,184
	Positioning, Navigation, and Timing	1,267
	ITS: Core Program - Multi-modal Research Applications	[39,733]
	ITS: Core Program - Multi-modal Research Technology	[8,940]
	ITS: Core Program - Multi-modal Research Policy	[4,000]
	ITS: Wireless Innovation and Infrastructure Initiative	[25,000]
Improved safety experience	BTS Program - Safety Data Analysis (secondary)	[556]
	RD&T Program - Research Coordination (secondary)	121
	RD&T Program - Knowledge Management System (secondary)	81
	RD&T Program - Education & Outreach (secondary)	121
	RD&T Program - Technology Transfer (secondary)	121
	BTS Program - Airline Transportation Statistics	[400]
Other	BTS Program - National Transportation Library (secondary)	[460]
ENVIRONMENTAL SUSTAINABILITY		
Reduced carbon/emissions and dependence on fossil fuels and improved energy efficiency	BTS Program - Travel Statistics (secondary)	[1,206]
	BTS Program - Freight Statistics (secondary)	[905]
	BTS Program - Geospatial Information statistics (secondary)	[181]
	BTS Program - Transportation Analysis, Data Quality and Performance Metrics (secondary)	[603]
	RD&T Program - Alternative Fuels (primary)	1,376
	RD&T Program - Research Coordination (primary)	91
	RD&T Program - Knowledge Management System (primary)	69
	RD&T Program - Education & Outreach (primary)	69
	RD&T Program - Technology Transfer (primary)	45
	Positioning, Navigation, and Timing (secondary)	317
Reduced pollution impacts on ecosystems	ITS: Core Program - Multi-modal Research Applications	[6,622]
	ITS: Wireless Innovation and Infrastructure Initiative	[25,000]
Environmentally sustainable practices and materials in transportation	BTS Program - Freight Statistics (secondary)	[301]
	RD&T Program - Research Coordination (secondary)	69
	RD&T Program - Knowledge Management System (secondary)	45
	RD&T Program - Education & Outreach (secondary)	69
	RD&T Program - Technology Transfer (secondary)	69
	BTS Program - Airline Transportation Statistics	[400]

	Exhibit II-3a	
	FY 2012 BUDGET REQUEST BY DOT OUTCOMES	
	RESEARCH AND INNOVATIVE TECHNOLOGY ADMINISTRATION	
	($000)	
DOT Outcome	**Program**	**FY 2012 Request**
Environmentally sustainable practices in DOT services and facilities		
Other	BTS Program - National Transportation Library	[350]
GOOD REPAIR		
	BTS Program - Travel Statistics (secondary)	[339]
	BTS Program - Freight Statistics (secondary)	[3,382]
	BTS Program - Geospatial Information Systems	[203]
	BTS Program - Transportation Analysis, Data Quality and Performance Metrics (secondary)	[1,082]
	BTS Program - National Transportation Library (secondary)	[392]
	RD&T Program - Research Coordination	323
	RD&T Program - Knowledge Management System	118
	RD&T Program - Education & Outreach	54
	BTS Program - Airline Transportation Statistics	[400]
LIVABLE COMMUNITIES		
Convenient and affordable choices	BTS Program - Travel Statistics (primary)	[955]
	BTS Program - Geospatial Information Systems (primary)	[397]
	RD&T Program - Research Coordination (primary)	157
	RD&T Program - Knowledge Management System (primary)	94
	RD&T Program - Education & Outreach (primary)	94
	RD&T Program - Technology Transfer (primary)	88
	BTS Program - Airline Transportation Statistics	[400]
	ITS: Core Program - Mode Specific Research	[4,500]
	ITS: Core Program - Other Legacy Initiatives	[2,704]
	ITS: Wireless Innovation and Infrastructure Initiative	[25,000]
Improved public transit experience	BTS Program - Travel Statistics (secondary)	[319]
Improved networks that accommodate pedestrians and bicycles		
Improved access for special needs populations		
Other	BTS Program - Freight Statistics (secondary)	[637]
	BTS Program - Transportation Analysis, Data Quality and Performance Metrics (secondary)	[956]
	BTS Program - National Transportation Library (secondary)	[369]
ECONOMIC COMPETITIVENESS		
Maximize economic returns	BTS Program - Transportation Economics (primary)	[1,745]
	BTS Program - Travel Statistics (secondary)	[978]
	BTS Program - Freight Statistics (primary)	[5,214]
	BTS Program - Geospatial Information Systems (secondary)	[196]
	BTS program - National Transportation Library (secondary)	[378]
	BTS Program - Airline Transportation Statistics	[360]

	Exhibit II-3a	
	FY 2012 BUDGET REQUEST BY DOT OUTCOMES **RESEARCH AND INNOVATIVE TECHNOLOGY ADMINISTRATION** **($000)**	
DOT Outcome	**Program**	**FY 2012 Request**
Competitive transportation system	BTS Program - Travel Statistics (secondary)	[326]
	BTS Program - Transportation Analysis, Data Quality and Performance Metrics (primary)	[1,303]
	RD&T Program - Research Coordination (primary)	59
	RD&T Program - Knowledge Management System (primary)	17
	RD&T Program - Education & Outreach (primary)	42
	RD&T Program - Technology Transfer (secondary)	67
	BTS Program - Airline Transportation Statistics	[2,280]
	ITS: Core Program - Multi-modal Research Applications	[19,866]
	ITS: Core Program - Multi-modal Research Technology	[2,235]
	ITS: Core Program - Exploratory Research	[2,000]
	ITS: Wireless Innovation and Infrastructure Initiative	[25,000]
Advance U.S. transportation interests abroad	BTS Program - Freight Statistics (secondary)	[1,303]
Expanded opportunities for businesses	RD&T Program - Research Coordination (secondary)	57
	RD&T Program - Knowledge Management System (secondary)	17
	RD&T Program - Education & Outreach (secondary)	41
	RD&T Program - Technology Transfer (primary)	101
	BTS Program - Airline Transportation Statistics	[360]
ORGANIZATIONAL EXCELLENCE		
	BTS Program - Travel Statistics (secondary)	[212]
	BTS Program - Freight Statistics (secondary)	[209]
	BTS Program - Transportation Analysis, Data Quality and Performance Metrics (secondary)	[321]
	BTS Program - National Transportation Library (primary)	[413]
	BTS Program - Airline Transportation Statistics	[400]
	ITS: Core Program - Technology Transfer & Evaluation	[13,900]
	ITS: Core Program - Program Support	[5,500]
OVERHEAD PROGRAMS/FUNCTIONS DISTRIBUTED TO PROGRAMS		
	Other R&D Salaries and Administrative Expenses	3,673
TOTAL		17,600

Notes:
* RD&T direct FTE are distributed based on actual FTE allocated to the program. Overhead FTE are distributed to Organizational Excellence.

* Airline Transportation Statistics FTE are distributed proportionately among each goal.

* BTS FTE are distributed proportionately among each goal.

EXHIBIT II-4

FY 2012 BUDGET AUTHORITY
RESEARCH AND INNOVATIVE TECHNOLOGY ADMINISTRATION
$(000)

	FY 2010 ACTUAL	FY 2011 CR (ANNUALIZED)	FY 2012 REQUEST
Research and Development	13,007	13,007	17,600
Bureau of Transportation Statistics			
Safety Data and Analysis	0	0	3,049
Travel Statistics	2,947	2,947	3,064
Freight Statistics	10,723	10,723	16,021
Transportation Economics	1,811	1,811	1,938
Geospatial Information	1,758	1,758	1,521
Transportation Analysis, Data Quality and Performance Metrics	7,416	7,416	7,268
National Transportation Library	2,345	2,345	2,139
Total - Bureau of Transportation Statistics [1/]	[27,000]	[27,000]	[35,000]
TOTAL: [Discretionary]	13,007	13,007	17,600

[1/] Resources are shown as non-adds because the Bureau of Transportation Statistics is an allocation account under the Federal-Aid Highway program.

EXHIBIT II-5

FY 2012 OUTLAYS
RESEARCH AND INNOVATIVE TECHNOLOGY ADMINISTRATION
($000)

ACCOUNT NAME	FY 2010 ACTUAL	FY 2011 CR (ANNUALIZED)	FY 2012 REQUEST
Research and Development	84,969	97,518	17,141
VOLPE National Transportation Systems Center	(23,438)	-	-
TOTAL	**61,531**	**97,518**	**17,141**

EXHIBIT II-6

SUMMARY OF REQUESTED FUNDING CHANGES FROM BASE
RESEARCH AND INNOVATIVE TECHNOLOGY ADMINISTRATION
RESEARCH AND DEVELOPMENT
Appropriations, Obligation Limitations, and Exempt Obligations
($000)

	2011 CR Annualized	Adjustments to Base	Annualization of 2011 FTE	Baseline Changes 2012 Pay Raise	One Less Compensable Day	GSA Rent	WCF Inc/Dec	Inflation/ Deflation	FY 2012 Baseline Estimate	Program Increases/ Decreases	FY 2012 Request
PERSONNEL RESOURCES											
Direct FTE	26.0	-							26.0		26.0
FINANCIAL RESOURCES:											
ADMINISTRATIVE EXPENSES											
Salaries and Benefits	4,172				-16			0	4,156	0	4,156
Travel	145							0	145	0	145
Training	40							0	40	0	40
GSA Rent	563					24		3	590	0	590
Printing	5							0	5	0	5
Other Services:											
-WCF	954						612	0	1,566	0	1,566
-Common Services	917							6	923	0	923
Equipment	150							0	150	0	150
Supplies	25							0	25	0	25
Admin Subtotal	6,971	0	0	0	-16	24	612	9	7,600	0	7,600
PROGRAMS											
Alternative Fuels R&D	500							0	500	0	500
RD&T Coordination	536							0	536	364	900
NDGPS	4,600							0	4,600	3,000	7,600
PNT	400							0	400	600	1,000
Programs Subtotal	6,036	0	0	0	0	0	0	0	6,036	3,964	10,000
TOTAL	13,007	0	0	0	-16	24	612	9	13,636	3,964	17,600

EXHIBIT II-6

SUMMARY OF REQUESTED FUNDING CHANGES FROM BASE
RESEARCH AND INNOVATIVE TECHNOLOGY ADMINISTRATION
Bureau of Transportation Statistics
Appropriations, Obligation Limitations, and Exempt Obligations
($000)

	2011 CR Annualized	Adjustments to Base	Baseline Changes					FY 2012 Baseline Estimate	Program Increases/ Decreases	FY 2012 Request
			Annualization of 2011 FTE	2012 Pay Raise	One Less Compensable Day	GSA Rent Inc/Dec	WCF Inflation/ Deflation			
PERSONNEL RESOURCES										
Direct FTE	70							70		70
FINANCIAL RESOURCES:										
ADMINISTRATIVE EXPENSES										
Salaries and Benefits	12,200				-47		0	12,153	0	12,153
Travel/Transportation of Things	205						1	206	0	206
Training	188						1	189	0	189
GSA Rent	1,443						7	1,450	0	1,450
Printing	10						0	10	0	10
Other Services:										
-WCF	3,553					107	0	3,660	0	3,660
-Common Services	2,398						12	2,410	0	2,410
Supplies	40						0	40	0	40
Equipment	750						4	754	0	754
Admin Subtotal	**20,787**	**0**	**0**	**0**	**-47**	**0**	**107**	**25**	**20,872**	**0**	**20,872**
PROGRAMS										
Safety Data and Analysis	0							0	2,000	2,000
Travel Statistics	450							450	0	450
Freight Statistics	4,500							4,500	5,915	10,415
Transportation Economics	160							160	0	160
Geospatial Information	360							360	0	360
Transportation Analysis, Data Quality and Performance Metrics	395							395	0	395
National Transportation Library	348							348	0	348
Programs Subtotal	**6,213**	**0**	**0**	**0**	**0**	**0**	**0**	**6,213**	**7,915**	**14,128**
TOTAL	**27,000**	**0**	**0**	**0**	**-47**	**0**	**107**	**25**	**27,085**	**7,915**	**35,000**

EXHIBIT II-7

WORKING CAPITAL FUND
RESEARCH AND INNOVATIVE TECHNOLOGY ADMINISTRATION
($000)

	FY 2010 ACTUAL	FY 2011 CR (ANNUALIZED)	FY 2012 REQUEST	CHANGE 2010 - 2012
Direct Account:				
Research and Development	928	954	1,566	638
Allocation Account:				
Bureau of Transportation Statistics	3,357	3,553	3,660	303
Reimbursable Account:				
Airline Transportation Statistics Program	873	795	929	56
Volpe National Transportation Systems Center	236	491	469	233
SUBTOTAL	1,109	1,286	1,398	289
TOTAL	**5,394**	**5,793**	**6,624**	**1,230**

EXHIBIT II-8

RESEARCH AND INNOVATIVE TECHNOLOGY ADMINISTRATION
PERSONNEL RESOURCE - SUMMARY
TOTAL FULL-TIME EQUIVALENTS

	FY 2010 ACTUAL	FY 2011 CR (ANNUALIZED)	FY 2012 REQUEST
DIRECT FUNDED BY APPROPRIATION			
Research and Development	26	26	26
REIMBURSABLE/ALLOCATION/OTHER:			
<u>Allocation Account:</u>			
Bureau of Transportation Statistics	68	70	70
<u>Reimbursable:</u>			
Airline Transportation Statistics Program	13	19	19
Transportation Safety Institute	32	41	41
Subtotal, Reimbursable Funded:	**45**	**60**	**60**
Volpe National Transportation Systems Center	532	532	532
<u>Other:</u>			
Intelligent Transportation Systems [non-add]	[18]	[18]	[18]
Subtotal, Reimbursable/Allocation/Other:	**645**	**662**	**662**
TOTAL FTEs	**671**	**688**	**688**

EXHIBIT II-9

RESEARCH AND INNOVATIVE TECHNOLOGY ADMINISTRATION
RESOURCE SUMMARY - STAFFING
FULL-TIME PERMANENT POSITIONS

	FY 2010 ACTUAL	FY 2011 CR (ANNUALIZED)	FY 2012 REQUEST
DIRECT FUNDED BY APPROPRIATION			
Research and Development	26	36	36
REIMBURSABLE/ALLOCATION/OTHER:			
Allocation Account:			
Bureau of Transportation Statistics	68	117	117
Reimbursable:			
Airline Transportation Statistics Program	13	19	19
Transportation Safety Institute	32	51	51
	45	70	70
Volpe National Transportation Systems Center	532	550	550
Other:			
Intelligent Transportation Systems [non-add]	[18]	[18]	[18]
Subtotal, Reimbursable/Allocation/Other:	645	737	737
TOTAL POSITIONS	671	773	773

SURFACE TRANSPORTATION AUTHORIZATION PROPOSAL
($000)

ACCOUNT NAME	FY 2012	FY 2013	FY 2014	FY 2015	FY 2016	FY 2017
Allocation/Reimbursable/Other Programs						
Competitive University Transportation Center (UTC) Consortia (FHWA)	72,000	72,000	72,000	72,000	72,000	72,000
University Transportation Center (UTC) Multimodal Competitive Research Grants	20,000	21,000	22,000	23,000	24,000	25,000
Multimodal Innovative Research Program	20,000	21,000	22,000	23,000	24,000	25,000
Intelligent Transportation Systems: Core Program	110,000	110,000	110,000	110,000	110,000	110,000
Total - Allocation/Reimbursable/Other	**222,000**	**224,000**	**226,000**	**228,000**	**230,000**	**232,000**
Bureau of Transportation Statistics						
Safety Data and Analysis	3,049	3,064	3,079	3,094	3,109	3,125
Travel Statistics	3,064	3,101	3,138	3,176	3,214	3,253
Freight Statistics	16,021	16,099	16,023	15,457	17,574	21,375
Transportation Economics	1,938	1,963	1,988	2,014	2,040	2,066
Geospatial Information Systems	1,521	1,537	1,553	1,570	1,587	1,604
Transportation Analysis, Data Quality and Performance Metrics	7,268	7,364	7,462	7,561	7,661	7,763
National Transportation Library	2,139	2,164	2,189	2,215	2,241	2,268
Total - Bureau of Transportation Statistics	**35,000**	**35,292**	**35,432**	**35,087**	**37,426**	**41,454**

RESEARCH AND INNOVATIVE TECHNOLOGY ADMINISTRATION

RESEARCH AND DEVELOPMENT

For necessary expenses of the Research and Innovative Technology Administration, [$17,200,000]*$17,600,000,* of which $10,000,000 shall remain available until September 30, [2013]*2014* : *Provided* , That there may be credited to this appropriation, to be available until expended, funds received from States, counties, municipalities, other public authorities, and private sources for expenses incurred for training. *Note.-- A full year 2011 appropriation for this account was not enacted at the time the budget was prepared; therefore, this account is operating under a continuing resolution (P.L. 111-242, as amended). The amounts included for 2011 reflect the annualized level provided by the continuing resolution.*

EXHIBIT III-1

RESEARCH AND DEVELOPMENT
SUMMARY BY PROGRAM ACTIVITY
Appropriations, Obligation Limitations, and Exempt Obligations
($000)

	FY 2010 ACTUAL	FY 2011 CR (ANNUALIZED)	FY 2012 REQUEST	CHANGE FY 2010-2012
Research and Development				
Salaries and Administrative Expenses	6,971	6,971	7,600	629
Alternative Fuels R&D	500	500	500	-
RD&T Coordination	536	536	900	364
Nationwide Differential Global Positioning System	4,600	4,600	7,600	3,000
Positioning, Navigation and Timing (PNT)	400	400	1,000	600
TOTAL: [Discretionary]	13,007	13,007	17,600	4,593
FTEs				
Direct Funded	26	26	26	0
Allocation/Reimbursable/Other:				
Transportation Safety Institute	32	41	41	9
Volpe National Transportation Systems Center	532	532	532	0
Intelligent Transportation Systems [non-add]	[18]	[18]	[18]	[0]

EXHIBIT III-1a

RESEARCH AND DEVELOPMENT
SUMMARY ANALYSIS OF CHANGE FROM FY 2011 TO FY 2012
Appropriations, Obligations, Limitations and Exempt Obligations
($000)

Item	Change from FY 2011 to FY 2012	FTE Change from FY 2011 to 2012
FY 2011 CR (ANNUALIZED)	13,007	26
RESEARCH AND DEVELOPMENT		
Adjustments to Base		
One Less Compensable Day	-16	
GSA Rent Increase	24	
Working Capital Fund Inc/Dec	612	
Inflation/Deflation	9	
Subtotal, Adjustments to Base	629	0
New or Expanded Programs		
RD&T Coordination	364	
Nationwide Differential Global Positioning System (NDGPS)	3,000	
Positioning, Navigation and Timing (PNT)	600	
Subtotal, New or Expanded Program Increases/Decreases	3,964	0
Total FY 2012 Request	17,600	26

EXHIBIT III-2
ANNUAL PERFORMANCE RESULTS AND TARGETS
Research & Innovative Technology Administration

The Office of Research, Development & Technology (RD&T) integrates performance results into its budget request to demonstrate alignment with the Department of Transportation's Strategic Plan. RD&T tracks the following agency measures in support of USDOT strategic goals of Environmental Sustainability, Economic Competitiveness, and Organizational Excellence:

USDOT Goal/Outcome: Environmental Sustainability

Alternative Fuels Program: Percentage of USDOT Operating Administrations research requests funded	2007	2008	2009	2010	2011	2012
Target		---	---	---	---	---	---
Actual		---	---	---	---	Establish baseline & target for FY 2012	

USDOT Goal/Outcome: Organizational Excellence

Research Coordination: Visits to Research Cluster Sharepoint sites at transportationresearch.gov by USDOT Researchers, USDOT-Funded Researchers, Stakeholders & Public	2007	2008	2009	2010	2011	2012
Target		---	---	---	---	500	1,000
Actual		---	---	---	---		

USDOT Goal/Outcome: Organizational Excellence

Research Coordination: Visits to Knowledge Management System by USDOT Personnel, Oversight Agencies, Congress & Public	2007	2008	2009	2010	2011	2012
Target		---	---	---	---	500	1,000
Actual		---	---	---	---		

USDOT Goal/Outcome: Organizational Excellence

Research Coordination: Percentage of USDOT RD&T Funded Projects Tracked by the Knowledge Management System	2007	2008	2009	2010	2011	2012
Target	---	---	---	---	70%	90%
Actual	---	---	---	---		

USDOT Goal/Outcome: Organizational Excellence

Research Coordination: Number of USDOT Funded Research Products and Applications Identified as Ready for Commercialization and/or Being Used in Our Transportation System and Posted on the RITA Website	2007	2008	2009	2010	2011	2012
Target	---	---	---	---	200	300
Actual	---	---	---	---		

The Nationwide Differential Global Positioning System (NDGPS) Program uses performance results for program management purposes and to demonstrate alignment with the Department of Transportation's Strategic Plan. NDGPS tracks the following agency measures in support of the USDOT strategic goal of Safety:

USDOT Goal/Outcome: Safety

Nationwide Differential GPS: System Availability (% of Broadcast GPS Corrections Available to Users; All Towers; Annualized)	2007	2008	2009	2010	2011	2012
Target	98.5%	98.5%	98.5%	98.5%	98.5%	98.5%
Actual	99.0%	98.9%	98.7%	98.6%		

* FY11 year to date observed

USDOT Goal/Outcome: Safety

Nationwide Differential GPS: Modernized Configuration (% of NDGPS Towers Brought to Current Coast Guard Operating Standards)	2007	2008	2009	2010	2011	2012
Target	---	---	---	---	10%	52%
Actual	---	---	---	---		

** Projected based on FY11 funding availability

The Positioning, Navigation, and Timing (PNT) program integrates performance results into its budget request to demonstrate alignment with the Department of Transportation's Strategic Plan. PNT tracks the following agency measures in support of the USDOT strategic goal of Safety:

USDOT Goal/Outcome: Safety

Positioning, Navigation, and Timing: Assessment of Progress on Implementation of National PNT Architecture*	2007	2008	2009	2010	2011	2012
Target	---	---	---	---	---	Baseline
Actual	---	---	---	---	---	

*Metric will be a green/yellow/red indicator on the progress of architecture implementation as measured by the PNT program.

RESEARCH AND INNOVATIVE TECHNOLOGY ADMINISTRATION
RESEARCH AND DEVELOPMENT
Program and Performance

The Research and Innovative Technology Administration (RITA) is responsible for coordinating, facilitating, and reviewing the Department's research and development programs and activities. Coordination and advancement of research and technology activities is led by the RITA Office of Research, Development and Technology and is funded through the General Fund. RITA is also responsible for coordinating and developing Positioning, Navigation and Timing (PNT) technology, PNT policy coordination, and spectrum management. RITA is the program manager for the Nationwide Differential Global Positioning System.

RITA oversees and provides direction to the following programs and activities:

The Bureau of Transportation Statistics (BTS) manages and shares statistical knowledge and information on the Nation's transportation systems, including statistics on freight movement, geospatial transportation information, and transportation economics. BTS is funded by an allocation from the Federal Highway Administration's Federal-Aid Highways account.

The Intelligent Transportation Systems (ITS) Joint Program Office (JPO) facilitates the deployment of technology to enhance the safety, efficiency, convenience, and environmental sustainability of surface transportation. The ITS program carries out its goals through research and development, operational testing, technology transfer, training and technical guidance. The ITS Research Program is currently funded through the Federal Highway Administration (FHWA). As part of the Wireless Innovation and Infrastructure Initiative, ITS will receive $100 million in mandatory resources in 2012, to conduct innovative wireless technology applications for transportation.

The University Transportation Centers (UTC) advance U.S. technology and expertise in many transportation-related disciplines through grants for transportation education, research, and technology transfer at university-based centers of excellence. The UTC Program funding is provided to RITA through an allocation from the Federal Highway Administration and a reimbursable agreement from the Federal Transit Administration.

The John A. Volpe National Transportation Systems Center (Cambridge, MA) provides expertise in research, analysis, technology deployment, and other technical knowledge to DOT and non-DOT customers on specific transportation system projects or issues, on a fee-for-service basis.

The Transportation Safety Institute develops and conducts safety, security, and environmental training, products, and services for both the public and private sector on a fee-for-service and tuition basis.

RESEARCH AND INNOVATIVE TECHNOLOGY ADMINISTRATION
RESEARCH AND DEVELOPMENT
PROGRAM AND FINANCING
(In thousands of dollars)

	Identification code 69-1730	FY 2010 ACTUAL	FY 2011 CR (ANNUALIZED)	FY 2012 REQUEST
	Obligations by program activity:			
0001	Salaries and administrative expenses	6,884	6,971	7,600
0002	Alternative fuels research & development	493	887	500
0003	Research development & technology coordination	575	794	900
0004	Nationwide differential global positioning system	4,600	4,600	7,600
0005	Positioning navigation & timing	400	400	1,000
0091	Direct program activities, subtotal	12,951	13,652	17,600
0100	**Direct program activities, Subtotal (running)**	12,951	13,652	17,600
0801	University transportation centers	3,095	7,600	8,000
0802	Transportation safety institute	10,043	20,000	20,000
0803	Other programs	6,963	10,000	10,000
0809	Reimbursable program activities, subtotal	20,102	37,600	38,000
0899	**Total reimbursable obligations**	20,102	37,600	38,000
0900	**Total new obligations**	33,053	51,252	55,600
	Budgetary resources:			
	Unobligated balance:			
1000	Unobligated balance brought forward, Oct 1	674	645	0
2200	New budget authority (gross)			
1021	Recoveries of prior year unpaid obligations	1,858	-	-
1050	Unobligated balance (total)	2,532	645	0
	Budget authority:			
	Appropriations, discretionary:			
1100	Appropriation	13,007	13,007	17,600
1160	Appropriation, discretionary (total)	13,007	13,007	17,600
	Spending authority from offsetting collections, discretionary:			
1700	Collected	24,498	37,600	38,000
1701	Change in uncollected payments, Federal sources	(6,246)		
1750	Spending auth from offsetting collections, disc (total)	18,251	37,600	38,000
1900	Budget authority (total)	31,258	50,607	55,600
1930	Total budgetary resources available	33,790	51,252	55,600
	Memorandum (non-add) entries:			
1941	Unexpired unobligated balance, end of year	645	0	0
1951	Unobligated balance expiring	(92)	-	-
	Change in obligated balance:			
	Obligated balance, start of year (net):			
3000	Unpaid obligations, brought forward, Oct 1 (gross)	173,885	107,369	23,504
3001	Adjustments to unpaid obligations, brought forward, Oct 1	10,636	-	-
3010	Uncollected pymts, Fed sources, brought forward, Oct 1	(16,321)	(22,203)	(22,203)
3011	Adjustments to uncollected pymts, Fed sources, brought forward, Oct 1	(10,635)		
3020	Obligated balance, start of year (net)	157,564	85,166	1,301
3030	Obligations incurred, unexpired accounts	33,053	51,252	55,600
3031	Obligations incurred, expired accounts	1,280		
3040	Outlays (gross)	(106,421)	(135,118)	(55,141)
3050	Change in uncollected pymts, Fed Sources, unexpired	6,246	-	-
3051	Change in uncollected pymts, Fed Sources, expired	(1,493)	-	-
3080	Recoveries of prior year unpaid obligations, unexpired	(1,858)	-	-
3081	Recoveries of prior year unpaid obligations, expired	(3,204)	-	-
	Obligated balance, end of year (net):			
3090	Unpaid obligations, end of year (gross)	107,369	23,504	23,963
3091	Uncollected pymts, Fed sources, end of year	(22,203)	(22,203)	(22,203)
3100	Obligated balance, end of year (net)	85,166	1,301	1,760
	Budget authority and outlays, net:			
	Discretionary:			
4000	Budget authority, gross	31,258	50,607	55,600
	Outlays, gross:			
4010	Outlays from new discretionary authority	21,990	49,306	53,840
4011	Outlays from discretionary balances	84,432	85,811	1,301
4020	Outlays, gross (total)	106,421	135,118	55,141
	Offsets against gross budget authority and outlays:			
	Offsetting collections (collected) from:			
4030	Federal sources	(21,453)	(37,600)	(38,000)
	Additional offsets against gross budget authority only:			
4050	Change in uncollected paymts, Fed sources, unexpired	(6,246)	-	-
4052	Offsetting collection credited to expired accounts	(3,201)	-	-
4060	Additional offsets against budget authority only (total)	3,045	-	-
4070	Budget authority, net (discretionary)	13,007	13,007	17,600
4080	Outlays, net (discretionary)	84,969	97,518	17,141
4180	Budget authority, net (total)	13,007	13,007	17,600
4190	Outlays, net (total)	84,969	97,518	17,141

RESEARCH AND INNOVATIVE TECHNOLOGY ADMINISTRATION
RESEARCH AND DEVELOPMENT
OBJECT CLASSIFICATION
(In thousands of dollars)

		FY 2010 ACTUAL	FY 2011 CR (ANNUALIZED)	FY 2012 REQUEST
	Direct Obligations:			
	Personnel compensation:			
1111	Personnel compensation, full-time permanent	2,484	3,098	3,087
1113	Other than full-time permanent	249	228	227
1115	Other personnel compensation	65	55	55
	Total personnel compensation	2,798	3,381	3,369
1121	Civilian personnel benefits	752	790	787
1210	Travel and transportation of persons	136	145	145
1220	Transportation of things	-	-	-
1231	Rent to GSA	487	563	590
1240	Printing and reproduction	-	5	5
1251	Advisory and assistance services	480	511	745
1252	Other services	450	550	4,050
1253	Other purchases of goods and services	7,540	7,532	7,734
1260	Office supplies	18	25	25
1310	Equipment	290	150	150
1990	Subtotal, direct obligations	12,952	13,653	17,600
2990	Reimbursable obligations	20,102	37,600	38,000
9999	Total obligations	33,053	51,253	55,600

Employment Summary:

		FY 2010 ACTUAL	FY 2011 CR (ANNUALIZED)	FY 2012 REQUEST
	Direct:			
1001	Civilian Full-time Equivalent Employment	26	26	26
	Reimbursable:			
2001	Civilian Full-time Equivalent Employment	45	60	60
	Allocation account:			
3001	Civilian Full-time Equivalent Employment	68	70	70

RESEARCH AND INNOVATIVE TECHNOLOGY ADMINISTRATION
RESEARCH AND DEVELOPMENT
Funding History
(In thousands of dollars)

YEAR	REQUEST	ENACTED
2005	0	4,213 [1]
2006	6,274	5,716 [2]
2007	8,217	7,736 [3]
2008	12,000	12,000 [4]
2009	12,000	12,900 [5]
2010	13,179	13,007 [6]
2011	17,200	-
2012	17,600	-

[1] Total FY 2005 enacted level for R&D was $5,967,369; $4,310,000 reflects the amount transferred to RITA. Previous funding for R&D appeared in the Research and Special Programs budget.

[2] FY 2006 reflects a 1% across the board rescission of $58,000 as stated in P.L. 109-148, section 3801.

[3] FY 2007 reflects Continuing Resolution (H.J. Resolution 20) at the FY 2006 budget level of $5,736,000 and fund to support Air Transportation Statistics program.

[4] FY 2008 reflects funding provided in P.L. 110-161.

[5] FY 2009 reflects funding provided in P.L. 111-8.

[6] FY 2010 reflects funding provided in P.L. 111-117.

Detailed Justification for Salaries and Administrative Expenses

What Do I Need to Know Before Reading This Justification?

- The core of RITA's mission is to coordinate research across the USDOT to maximize and leverage the taxpayers' $1.2 billion annual investment in transportation research, development and technology (RD&T) activities.
- The "Salaries and Administrative Expenses" line item provides resources for the staff funded by the RITA General Fund account. These staff support "the coordination, facilitation, and review of the Department's research and development programs and activities" as originally described in the Norman Y. Mineta Research and Special Programs Improvement Act.

What Is the Request and What Will We Get for The Funds?

FY 2012 Salaries and Administrative Expenses Budget Request ($000)

Program Activity	FY 2010 Actual	FY 2012 Request	Change FY 2010-2012
Salaries and Administrative Expenses	6,971	7,600	629
Total	**$6,971**	**$7,600**	**$629**

The salaries and administrative expenses program provides for 26 FTE. The staff primarily supports the Research, Development and Technology (RD&T) program and provides administrative support necessary to implement RITA's coordination mission. In addition, these funds support administrative expenses including: travel, training, rent, working capital, IT support and E-government initiatives.

The funding increase of $.629M funds increases to rent and the working capital fund.

What is This Program?

Strategic Goals: Safety, State of Good Repair, Economic Competitiveness, Livable Communities, Environmental Sustainability and Organizational Excellence.

Detailed Justification for Alternative Fuels R&D

What Do I Need to Know Before Reading This Justification?

- The program continues to cover its expanded focus of addressing transportation issues for all alternative fuels, not just hydrogen, in the scope of its research and education topics.
- The program continues to convene USDOT program managers, researchers and external stakeholders to develop intermodal/multimodal projects, the results of which are shared broadly and support USDOT strategic goals, particularly environmental sustainability. The program also facilitates cooperation across the Administration on biofuels and with the Administration's interdepartmental Biomass Research and Development (R&D) Board.
- The program produces collaborative education products including distance learning tools to train emergency responders about the hazards of in-transit hydrogen incidents; distance and on-site learning for commercial motor vehicle safety inspectors. Pipeline innovations developed under the program can reduce or eliminate embrittlement of the infrastructure. The steel innovations developed under this program have the potential to improve the integrity of steel bridges and reduce the time required for construction, further leveraging the benefits of these research investments.

What Is the Request and What Will We Get for The Funds?

FY2012 – Alternative Fuels R&D Budget Request ($000)

Program Activity	FY 2010 Actual	FY 2012 Request	Change FY 2010-2012
Alternative Fuels R&D	500	500	0
Total	**$500**	**$500**	**0**

The request continues the base program at prior levels and continues its mandate to include all alternative fuels.

Key Actions: Activities described below are consistent with RITA's mission to coordinate cross-modal research. Funding supports stakeholder needs in alternative fuels and the USDOT's environmental sustainability strategic goal:

- Funding will support refinements and improvements to Pipeline and Hazardous Materials Safety Administration (PHMSA) emergency response training tools and Federal Motor Carrier Safety Administration (FMCSA) commercial vehicle safety training and education products that are adaptable for all fuels.
- Funding will support the DOT Alternative Fuels Working Group, including PHMSA, the National Highway Traffic Safety Administration (NHTSA), FMCSA and other U.S. DOT modal administrations, to conduct advanced research.
- Funding will support a competition to identify alternative fuels projects that improve the transportation system's ability to support this emerging industry and promote energy independence.

- Funding will continue to support collaboration with other federal agencies and external partners, to heighten awareness and leverage resources in conducting research on alternative fuels.
- Funding will support the adaptation and/or application to transportation of long-term enabling research performed by other Federal agencies, including the Departments of Energy and Agriculture, the National Science Foundation, and the Biomass R&D Board.
- Based on the principles of peer review and under the auspices of the multimodal RD&T Planning Council, a panel of USDOT program managers, researchers, and other experts will competitively allocate funds to intermodal and multimodal alternative fuel projects that advance development of codes and standards, emergency response and safety training as well as exploring technologies to eliminate pipeline embrittlement and improve pipeline-welding techniques.

Key Outputs: Multi-modal focused alternative fuels research and education products examining safety, logistical and distribution issues of alternative fuels approaching market level production. Standards, first responder training and safety handling products will assist in developing a responsive transportation system for these new fuels.

Key Outcomes: Safe and efficient transportation system components that accommodate alternative fuels through multi-modal absorption of the increased shipping/transport requirements of new fuels and feedstocks. Improved integration of Administration biofuel investments, especially with the Departments of Energy and Agriculture, the National Science Foundation, and the Administration's Biomass R&D Board.

What is This Program?

Strategic Goals: Environmental Sustainability.

Purpose/Beneficiaries: This program convenes the U.S. DOT alternative fuels program managers and researchers to select projects that leverage and improve the current state of knowledge.

Partners: FHWA, FAA, PHMSA, MARAD, FMCSA, FTA, FRA, NHTSA, OST

Description: Funding allows the modal administrations to bring together partners within the USDOT, from other federal agencies and external stakeholders to pursue activities with broad appeal and need. The program's competitive review process will allow the program to support initiatives that best support the collective transportation missions and objectives.

Prior Year Accomplishments:
- PHMSA emergency response training and FMCSA commercial vehicle safety training created a distance learning framework adaptable for all alternative fuels.
- PHMSA research developed pipeline standards for safe shipment of hydrogen by pipeline and will inform biofuels standards development.
- Projects delivered outputs that leveraged DOT research and education aimed at further development and deployment of alternative fuels.

Why Is This Program Necessary?

The Alternative Fuels program funds research and education aimed at reducing our reliance on petroleum as the primary transportation energy source. Alternative fuels research supports the establishment of safety standards for the use and conveyance of alternative fuels that are currently under development.

The program encourages and supports cooperation and leverages investments across the federal government to maximize the benefits of alternative fuels on our nation's transportation system Alternative fuel program managers and researchers cooperate across the USDOT and with stakeholders on topics such as the development of codes and standards and emergency response training. The program also supports the USDOT environmental sustainability strategic goal, and aims to expand the safe and efficient deployment of alternative fuels.

Without this funding, RITA's ability to support the deployment of alternative fuels and the opportunity for the public to choose between energy sources would be lost.

How Do You Know the Program Works?

Effective: The USDOT modal administrations and their stakeholders continue to develop proposals for consideration by the Alternative Fuels Working Group. The resulting research and education products continue to be well received and widely disseminated. In addition to information sharing and development of best practices, the results of this research will be measured in:
- Technological advances evidenced through patents and technology demonstrations.
- Increased scientific knowledge measured by numbers of publications and citations.
- Promulgation of training materials, rules, regulations and standards based on research results.

Research: Research to advance safe and efficient shipment of alternative fuels in a multi-modal system is needed to transport newly developing fuels from source to user. Current systems are structured for the chemical composition of petrochemical fuels, which present different corrosive, safety and flammability properties from alternative fuels and from feedstock used to create them. Standards, regulations, handling and training information is needed from research results.

Efficient: RITA's Office of RD&T plays a coordinating role, leveraging limited funds to pursue research that has multimodal relevance, particularly in the identification of safety issues that cross transportation modes, appropriate responses, corrosive properties and materials impacts.

Why Do We Want/Need To Fund The Program At The Requested Level?

Funding Options: The funding request is the minimum level needed to facilitate cooperation and leverage investment across the USDOT to address the impacts of alternative fuels on the transportation systems and to advance their safe and efficient deployment.

History/Outyear Needs: Funding has been consistent at $.500M since inception of the program. FY 2012 funds are requested at the same level with broadened research focus to include transportation issues for multiple alternative fuels.

Detailed Justification for RD&T Coordination

What Do I Need to Know Before Reading This Justification?

The core of RITA's mission is to coordinate research across the USDOT to maximize and leverage the taxpayers' $1.2B annual investment in transportation research, development and technology (RD&T) activities and to ensure that decisions are made based on sound science and rigorous analysis.

- RITA coordinates RD&T through the RD&T Planning Council, the RD&T Planning Team and DOT's research clusters (www.transportationresearch.gov) as well as through inter-departmental education events, coordination of technology transfer activities and outreach.
- Through the transportation Knowledge Management System (KMS), RITA provides transparency as well as information needed to identify opportunities for greater coordination and collaboration.

What Is the Request and What Will We Get for The Funds?

FY 2012 RD&T Coordination Budget Request ($000)

Program Activity	FY 2010 Actual	FY 2012 Request	Change FY 2010-2012
Education and Outreach	50	50	0
Knowledge System Enhancement & Maintenance	200	200	0
Technology Transfer Program Support	50	50	0
Coordination	236	600	364
Total	**$536**	**$900**	**$364**

Key Actions:
- Develop a Communication, Education and Outreach program using multiple Web 2.0 tools to disseminate information about transportation RD&T within the USDOT research community, increase awareness of USDOT funded university research and increase transparency to external stakeholders and research partners.
- Enhance and maintain the USDOT KMS database to provide better data mining and reporting and support its continued maintenance. The KMS will allow internal and external stakeholders to access and search for research in progress and research products in their areas of interest. This will help inform their research, avoid duplication, indentify individual researchers and program managers for collaboration, support policy decisions, and identify potential areas of collaboration.

- Disseminate research products and results to better facilitate their commercialization, technology transfer, and implementation.

What is the Program?

Strategic Goal: Safety, State of Good Repair, Economic Competitiveness, Livable Communities, and Environmental Sustainability.

Purpose/Beneficiaries: Provides strategic direction and fosters coordination and collaboration of the USDOT research programs. RD&T Coordination enhances the USDOT's ability to foster technology transfer, ensure implementation of research results and inform RD&T stakeholders of USDOT research activities.

Beneficiaries: USDOT cross-modal and mode-specific research programs, researchers both with within and outside of the USDOT and ultimately users and managers of the transportation system.

Partners: USDOT modal administrations, including members of the RD&T Planning Team (Associate Administrators for RD&T) develop the DOT RD&T Strategic Plan with input from a broad array of external stakeholders and partners. The RD&T Strategic Plan will be implemented and continuously updated and refined over the next five years and beyond.

Description: Databases containing cross-modal and mode-specific research program and project details are the bases of the overall KMS database and U.S. DOT research clusters supported by the www.transportationresearch.gov collaborative website used to facilitate collaboration and coordination. USDOT modal administrations are the sources of information used by RITA to inform other USDOT, university, state, and foundation researchers of our efforts—fostering coordination and collaboration.

Prior Year Accomplishments:
- Increased RD&T coordination through:
    - Monthly meetings of RD&T Planning Team
    - Convened 14 RD&T Research Clusters and held three rounds of cross-modal meetings among members.
    - Collaborative development of a draft RD&T Strategic Plan including external stakeholder input.
    - Establishment of USDOT Research Cluster areas and identification of USDOT program managers/researchers in priority areas.
    - Establishment of Research Cluster transportationresearch.gov collaborative/interactive websites for posting and dissemination of information on completed, current and planned research projects, people and products.
    - Brought together DOT modal administrations to drive collaborative research and analysis in support of the Secretary's Distracted Driving initiative.

- Communication activities included:
    - Quarterly site visits by RD&T Planning Team to USDOT modal administration research facilities to familiarize members with research capabilities and pursuits of modal partners.
    - Monthly publication of University Transportation Center (UTC) Program Spotlight newsletters, seminars, lectures and monthly RD&T Interchange email newsletters to all USDOT researchers, RD&T Planning Team members, UTC directors and other interested stakeholders.
    - External RITA RD&T Program website with added links to U.S. DOT research information.
    - Technical support to USDOT Climate Change Center lecture series.
    - Monthly meetings of the RD&T Planning Team and regularly scheduled meetings of research Cluster members.
    - The annual UTC "Spotlight Conference cosponsored with the Transportation Research Board (TRB).
- Completed basic functionality of RD&T Knowledge Management System (KMS) that contains searchable descriptors of project level activities in RD&T across the U.S. DOT and established staff position to manage the KMS database.
- Dedicated staff to develop and coordinate technology transfer program. RITA is gathering information and selecting best practices used at other agencies. The technology transfer program will support dissemination of information on research products for implementation and for commercial developments.

Why Is This Program Necessary?

RD&T coordination activities enable RITA to better recognize and use the synergies in the USDOT's RD&T programs and to identify opportunities for collaborative research efforts. Through the RD&T Planning Council and RD&T Planning Team the USDOT RD&T Strategic Plan is continually updated with external stakeholder input. The budget guidance, budget reviews and program reviews result in better coordinated transportation research, adoption of best practices in research program management and peer review, and minimize duplication. RD&T coordination includes the creation of USDOT research clusters made up of research program managers and researchers that meet regularly and interact via the transportationresearch.gov collaborative website. This encourages research program managers and researchers to interact, share planned and in-progress research plans, post implementable research results, conduct online collaboration, debate research questions, and report on research outcomes. These mechanisms will inform others on progress, topics being explored and products that can be leveraged. RD&T coordination also includes sponsorship of seminars and lectures by transportation researchers, particularly on multi-modal topics.

RD&T coordination also includes the KMS database that supports research coordination, facilitation, review, and strategic planning. The KMS database also fosters research coordination and collaboration across the USDOT and with stakeholders, and promotes excellence in research management, transparency and dissemination of readily implementable research results.

Beneficiaries include the USDOT, transportation research program managers and researchers, external stakeholders, transportation system managers and users, OMB, Congress.

Most transportation research is funded on a mode-by-mode basis. The RD&T Coordination program provides a mechanism for mode-specific research to be coordinated through the RD&T Planning Team and Planning Council. The advancements in information flow and awareness would stop. Duplication of effort, and a slowing in the adoption of best practices in research program management would likely occur. Leveraging and advancement of cross-modal USDOT Research Clusters would halt, and the USDOT KMS would not be available for coordinating research across transportation modes.

The program staff is dedicated to looking across the USDOT modal administrations to identify areas of potential synergy, and prevent overlap, and inefficiencies resulting from duplication of effort. Funding for a technology transfer program provides structured dissemination of information on research products across the USDOT and with stakeholders, reducing the potential for leveraging of funds and results and adoption of technologies across the transportation industry.

How Do You Know the Program Works?

Effective: Evidence received through the RD&T Planning Council and the RD&T Planning Team and during site visits by Planning Team members to modal administration research facilities demonstrate that effectiveness of the program has increased as a result of initiatives such as the multimodal coordination and development of the USDOT RD&T Strategic Plan and initiation of USDOT Research Clusters and transportationresearch.gov collaborative web site.

Research: RITA's RD&T coordination program is developing effective qualitative and quantitative metrics for these activities. Metrics in development are:
- Use of transportationresearch.gov collaborative website to seek, post and share information about research in progress, people and products.
- Number of collaborative research projects undertaken by USDOT.
- Number of patent applications posted and implementable research products listed on the transportationresearch.gov collaborative website.
- Number of projects, studies and other outputs shared across the USDOT as well as with federal partners and external stakeholders.
- Full representation of USDOT's $1.2B research portfolio within the KMS database.
- Use of the KMS to identify research gaps, overlaps, and potential synergies.

Efficient: Approaches developed and being pursued, to include the USDOT RD&T Strategic Plan, the Research Clusters, the collaborative transportationresearch.gov and the Knowledge Management System have all been endorsed and/or identified by the USDOT RD&T Planning Team comprising the modal Associate Administrators, and have been developed with their input for optimum utility and applicability.

Why Do We Want/Need To Fund The Program At The Requested Level?

Funding Options: The current funding level will allow the program to proceed as planned. This year's activities include:
- Maintain the Knowledge Management System (KMS), effectively identify and incorporate new research efforts into the database, upgrade software, and improve the system's functionality and reporting capabilities so that it maintains its usefulness as a USDOT-wide resource.
- Leverage the KMS to produce topic-related reports in key areas of interest.
- Enhance the strategic collaboration process for transportation RD&T.
- Identify products and outcomes resulting from USDOT RD&T investments through the technology transfer program and capitalize on events and activities sponsored by other Departments (e.g. DoD, USDA and NIST) such as the annual "World's Best Technology" showcase inviting researchers to present concepts to industry and investors.
- Increase USDOT participation in initiatives of the Office of Science and Technology Policy (OSTP)/National Science and Technology Council.
- Strengthen existing public/private technology partnerships and promote new collaborations through outreach to the entire transportation community.
- Conduct education programs and seminars, and support internships and sabbaticals to inform researchers within the USDOT, and funded universities, of the USDOT's ongoing RD&T activities.

Funding Options: FY 2012 requested funding of $.900M will support Knowledge Management System (KMS) database development and maintenance, education and outreach and support of researcher connectivity through meetings, webinars, and information dissemination.

History/Outyear Needs: FY 2010 $.536M; FY 2012 $.900M.

Detailed Justification for the Nationwide Differential Global Positioning System (NDGPS) Program

What Do I Need To Know Before Reading This Justification?

The Nationwide Differential Global Positioning System (NDGPS) Program provides national, real-time accurate dynamic positioning and navigation information at one-to-three meters for surface transportation users. NDGPS receives substantial routine use by highway, transit and rail sectors in operations, maintenance, planning, construction, quality assurance and asset management. Multiple other Federal agencies, state and local governments and private sector users in many economic sectors meet mission requirements in precision agriculture, surveying and mapping, GIS, environmental and natural resource management, and severe weather forecasting using the NDGPS service.

- The funding requested in the FY 2012 President's Budget aims to address system technical, cost and performance threats caused by deferred maintenance, by:
    - Establishing a robust operations and maintenance (O&M) floor for reliable system availability to users; and
    - Initiating a recapitalization of system equipment to bring the DOT sites up to current Coast Guard operational standards. The FY 2012 funding request provides 100 percent of routine O&M needs and 50 percent of recapitalization costs.
- NDGPS is funded through DOT to support surface transportation users. However, it is recognized that multiple Federal agencies/sub-agencies, state and local governments, and private sector users in many economic sectors leverage DOT's investment in NDGPS to meet their requirements.
- Although NDGPS is funded through RITA, NDGPS is an operational nationwide positioning and navigation utility, not a research and technology program. It provides an infrastructure to support limited DOT R&D projects.

What Is The Request And What Will We Get For The Funds?

FY 2012 – Nationwide Differential Global Positioning System (NDGPS) Budget Request ($000)

Program Activity	FY 2010 Actual	FY 2012 Request	Change FY 2010-2012
Nationwide DGPS	4,600	7,600	3,000
Total	$4,600	$7,600	$3,000

The request is to enable:
- NDGPS to maintain system availability at the 29 inland segment sites at 98.5% or better;
- Reduced risk of failure by bringing inland segment stations up to Coast Guard standards.

Key Actions	Key Outputs	Key Outcomes
• Exercise robust O&M protocols • Reduce maintenance backlog • Initiate equipment recapitalization effort	• Maintain system availability at ≥ 98.5% • Reduce programmatic risk (technical, cost, performance) • Eliminate outdated equipment	• Technology providers and users can make investments based on improved system reliability/reduced risks • Enables ability to troubleshoot issues remotely; outyear O&M cost growth reduced

What Is This Program?

Strategic Goals: The Nationwide Differential Global Positioning System (NDGPS) supports the safety and environmental sustainability strategic goals.

Purpose/Beneficiaries: Provides a national service that delivers real-time, accurate dynamic positioning and navigation information at one-to-three meters (and often submeter) for civil transportation applications. NDGPS accomplishes this goal through local broadcast of accurate GPS position corrections (http://www.navcen.uscg.gov/ndgps/default.htm). The Coast Guard operates NDGPS as a nationwide DGPS utility coordinated with the Maritime DGPS and inland waterways sites (http://www.navcen.uscg.gov/dgps/default.htm). DOT provides NDGPS for surface transportation applications and there is significant public and private sector reliance on NDGPS for precision agriculture, environmental management, natural resource management, surveying and mapping, severe weather forecasting and other applications.

Partners: U.S. Coast Guard and other federal, state and local agencies.

Description: NDGPS provides support infrastructure for several DOT research projects, and opportunities for future cross-modal surface transportation safety, security, efficiency and emergency response services.

Why Is This Particular Program Necessary?

NDGPS augments GPS by providing increased accuracy and integrity for transportation applications. (http://www.navcen.uscg.gov/AccessASP/NDgpsReportAllSites.asp). The NDGPS service provides reliable GPS to meet the growing requirements of surface transportation users.

In addition to providing a real-time broadcast of corrections, NDGPS provides a robust operational backbone to the DOC's CORS application (http://www.ngs.noaa.gov/CORS/) for post-processing survey applications and Web-enabled location solutions, the National Weather Service's Forecast Systems Laboratory for short-term severe weather and precipitation forecasts, space weather forecasts, and for plate tectonic monitoring.

The equipment recapitalization will bring the inland NDGPS segment to the Coast Guard operational standard, enabling remote troubleshooting and reducing outyear O&M cost growth. Coast Guard completed maritime DGPS recapitalization in 2009, and has realized significant operational efficiencies. Inland NDGPS equipment beyond serviceable life is being maintained at high risk of failure and at increased expense by purchasing/stocking/rebuilding expensive parts.

NDGPS availability and accuracy is leveraged by multiple other Federal agencies to meet their requirements in surveying and mapping, precision agriculture, environmental and natural resources management; historical resources management; and emergency response. A new use in FY 2011 was integration of the NDGPS in the Department of Homeland Security's (DHS) GPS Interference Detection and Monitoring (IDM) system for identifying GPS interference and service interruptions.

Without this funding, hundreds of thousands of users across multiple economic sectors would be unable to receive the accurate positioning information on which they depend daily for routine operations, and the growing location-based services market would be negatively impacted, as documented in "NDGPS Assessment: Final Report."

How Do You Know The Program Works?

Effective: DOT's "NDGPS Assessment: Final Report" documents hundreds of thousands of routine NDGPS users across multiple sectors, and a minimum system return on investment (ROI) for the Federal-Aid Highway Program of 40:1. ROI for other Federal agencies, state and local governments and private sector could not be calculated, but includes a significantly larger user base.

Research: Ongoing feedback received from the user community through the Civil GPS Service Interface Committee (CGSIC; http://www.navcen.uscg.gov/cgsic/), the worldwide forum for effective interaction between all civil GPS users, supports the continued need for the NDGPS service, and to the expectation of expanded future uses in transportation and other economic sectors, especially if accuracy can be improved.

Research: Recognizing the cross-sector interest in a robust, higher-accuracy augmentation, the U.S. Department of Agriculture is conducting a study suggesting expanded NDGPS applications, "Evaluating the Potential for New Civil User Services That Could Be Provided From DGPS Sites."

Efficient: New receiver manufacturers are producing commercial DGPS-based products for the market, demonstrating both market/user demand and a recognition of system reliability, and increasing cross-sector interest in a higher-accuracy augmentation.

Why Do We Want/Need To Fund The Program At The Requested Level?

Funding options: The requested level of funding is needed to support robust O&M to maintain system reliability and availability commitments to the Federal, state and local transportation user communities; and to reduce program technical and cost risk by continuing to replace aged equipment that is at risk of failure.

History/Outyear Needs: FY 2010 $4.6M; FY 2012 $7.6M.

Detailed Justification for the Positioning, Navigation, and Timing (PNT) Program

What Do I Need To Know Before Reading This Justification?

The Positioning, Navigation and Timing (PNT) Program is the sole means by which the Federal Government defines and implements civil sector PNT requirements. The Secretary of Transportation is assigned this role by National Security Presidential Directive, and includes all Federal, state and local government needs for GPS and its augmentations, as well as those of the private sector. The PNT Program is essential to ensuring that critical infrastructures have the primary and back-up PNT systems upon which they depend for daily operations, as well as identifying and pursuing gaps and research needed to meet these requirements, to enable future systems such as Next Gen, Positive Train Control, and Intelligent Transportation Systems (ITS).

DOT and DoD, in conjunction with 29 other departments and agencies, recently completed the National PNT Architecture Implementation Plan which identifies over 40 tasks to be accomplished to overcome capability gaps predominantly resulting from the limitations of GPS.

What Is The Request And What Will We Get For The Funds?

FY 2012 – Positioning, Navigation, and Timing Program (PNT)
Budget Request
($000)

Program Activity	FY 2010 Actual	FY 2012 Request	Change FY 2010-2012
PNT Policy Coordination	115	300	185
National PNT Architecture	275	650	375
Civil GPS Service Interface Committee	10	50	40
Total	$400	$1,000	$600

The request is for $1M to enable DOT to serve as the lead for PNT requirements, architecture development, and GPS acquisition, development, and operations for all United States Government civil departments and agencies.

This program produces the Federal Radionavigation Plan, Civil PNT Requirements Document, chairs the Civil GPS Service Interface Committee, and advances the National PNT Architecture in conjunction with the Department of Defense and other government agencies to close identified PNT capability gaps and provide more efficient and effective capabilities.

Key Actions	Key Outputs	Key Outcomes
• Hold meetings of the Federal Radionavigation Plan (FRP) Working Group to finalize 2012 FRP content • Coordinate 2012 FRP through DOT Extended Pos/Nav Working Group and Executive Committee • Coordinate update to Civil PNT Requirements Document with DOT Extended Pos/Nav Working Group • Demonstrate ability to close identified PNT capability gaps through implementation of the 2025 National PNT Architecture • Hold U.S. and international CGSIC meetings	• Publish the 2012 *Federal Radionavigation Plan* in conjunction with DoD and DHS • Produce updated Civil PNT Requirements Document • Provide a refined National PNT Architecture Implementation Roadmap based on coordinated analysis • Provide Summary Record of CGSIC meetings on website	• Transportation industry and technology providers can make business decisions on products to offer (or not offer) based on U.S. Government plans for PNT capabilities and services • Informed U.S. government and commercial decision making and ability to support long range planning on alternative PNT sources that can be integrated with GPS • Public knowledge of the status of GPS modernization and the ability for it to satisfy civil PNT requirements

What Is This Program?

Strategic Goals: Safety and environmental sustainability.

Purpose/Beneficiaries: The purpose of this effort is to provide more effective and efficient PNT capabilities, and an evolutionary path for government-provided PNT systems and services. The beneficiaries of this effort are civil departments and agencies in the development, acquisition, management, and operations of GPS and other PNT services, and external users of government-provided PNT services.

Partners: RITA serves as the civil lead of the National PNT Architecture effort, a cross-modal interagency effort, to guide future PNT system-of-systems investment and implementation decisions. This effort is conducted through the DOT Pos/Nav Working Group and Executive Committee to coordinate with all modes within DOT. The DOT Extended Pos/Nav Working Group and Executive Committee is DOT's PNT coordination process with other civil agencies. The structure for this coordination process is shown in Figure 1.

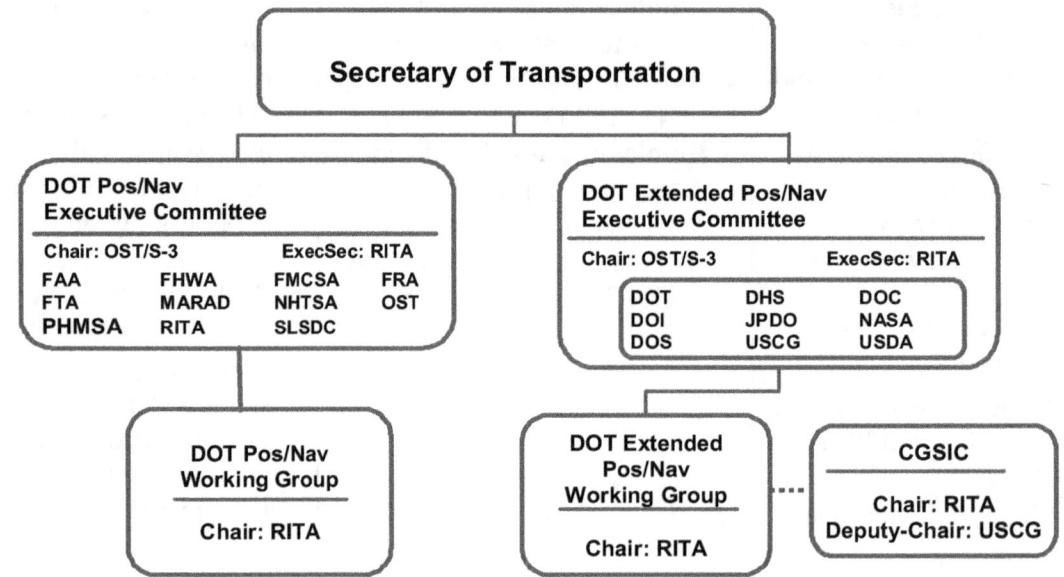

Figure 1 Civil PNT Coordination Structure

Description: The Positioning, Navigation, and Timing Program coordinates DOT PNT technology, policy, as well as provides civil PNT systems analysis which is critical to intermodal transportation applications in support of the DOT Strategic Goals. The program coordinates all civil sector PNT requirements and provides outreach to communities associated with PNT plans and policy.

A general description of the work of the program is provided below:
PNT Policy Coordination
- Provide support to the Deputy Secretary of Transportation and Under Secretary of Transportation for Policy with preparation of materials and briefings for their participation as the co-chairs of the Space-Based PNT Executive Committee and Executive Steering Group, respectively.
- Develop the *Federal Radionavigation Plan* (FRP) as directed by the National Defense Authorization Act for Fiscal Year 1998 (10 U.S.C. 2281(c)). The FRP is developed biennially by DOT in conjunction with DoD and DHS and is signed by the Secretaries of DOT, DoD, and DHS. The FRP reflects national policies and plans for U.S. government radionavigation systems and is used as a reference document both nationally and internationally.
- Provide a Civil PNT Requirements Document which serves as the foundation for the National PNT Architecture effort and allows program assessments to be performed to

determine if the capabilities of the National PNT Architecture meet the needs of the civil community.

National PNT Architecture
- The National PNT Architecture is based on the fact that PNT is integral to the infrastructure on which the U.S. economy, national security and transportation systems rely. The architecture provides a time-phased, fiscally informed roadmap through 2025 of PNT activities, including needs for research and development.

Civil GPS Service Interface Committee (CGSIC)
- The CGSIC is the recognized worldwide forum for effective interaction between all civil GPS users and the U.S. GPS authorities. RITA chairs the CGSIC and the Coast Guard serves as Deputy Chair.

Prior Year Accomplishments:
- Address the issue of emerging low-cost GPS jammers from both a technical and regulatory standpoint. Work with the Department of Homeland Security (DHS) on a GPS threat assessment for transportation applications of GPS, as well as a National Risk Assessment for applications of GPS.
- Release of the 2011 DOT Spectrum Plan. RITA will also participate in the Interagency Radio Advisory Committee and represent the DOT position in proceedings involving other agencies such as the Federal Communications Commission and the National Telecommunications and Information Administration of the Department of Commerce.
- Development of the draft of 2012 Federal Radionavigation Plan.
- Assessment of research and develop efforts as identified in the National PNT Architecture Plan.

Why Is This Particular Program Necessary?

The National Security Presidential Directive on Space-Based Positioning Navigation and Timing Policy gives the Secretary of Transportation broad responsibilities in providing for and implementing PNT services for the civil community. This policy ensures that the U.S. maintains space-based positioning, navigation, and timing services, augmentation, back-up, and complementary capabilities. Positioning, Navigation, and Timing capabilities are critical for the Next Generation Air Transportation System, Intelligent Transportation Systems, and Positive Train Control. In 2001, the Volpe Center released a study entitled *"Vulnerability Assessment of the Transportation Infrastructure Relying on the Global Positioning System"* http://ntl.bts.gov/lib/31000/31300/31379/17_2001_Volpe_GPS_Vulnerability_Study.pdf
In the nine years since that study was released, little has been done to implement solutions that address these vulnerabilities.

DOT, as the lead civil agency for PNT, has the opportunity to continue to partner with the Department of Defense to advance research on alternative PNT technologies that can be integrated with GPS to meet multiple user needs as identified in the National PNT Architecture Implementation Plan. This allows for long range planning through 2025 and the ability to make informed decisions on PNT capabilities based on solid technical and cost information.

How Do You Know the Program Works?

Before the National PNT Architecture effort was formed, the U.S. Government had an *ad hoc* approach to developing and providing PNT capabilities. There was no coordinated research, duplication of efforts, and capability gaps that no agency was addressing. There also was little recognition of the synergies of PNT needs across civil, military, and homeland security communities. These are identified the National PNT Architecture Study Report from 2008. http://ntl.bts.gov/lib/31000/31300/31341/06_2008_National_PNT_Architecture_Study_Final_Report.pdf.

The National PNT Architecture effort works because it recognizes that PNT is integral to the infrastructure on which the U.S. economy, national security and transportation systems rely. DOT and DOD, in conjunction with 29 other departments and agencies, recently completed the National PNT Architecture Implementation Plan which identifies tasks to be accomplished to overcome capability gaps resulting from the limitations of GPS.

Why Do We Want/Need To Fund The Program At The Requested Level?

USDOT requires funds at the requested level to fulfill its responsibilities as the lead civil agency for PNT. Technical and cost analysis is needed for the Department to make informed decisions on alternative sources of PNT to integrate with GPS.

History of PNT Funding and Out Year Needs: FY 2010 $.400M; FY 2012 $1M.

Research and Development
Allocation/Reimbursable/Other Programs

Competitive University Transportation Centers (UTC) Program	TOTAL FY 2012 Request: [$100,000,000]

The fully competitive University Transportation Center (UTC) program will consist of two components: a base portion of multimodal and multidisciplinary UTC consortia around particular theme areas plus an additional $20 million that will be reserved for a targeted multimodal research program for which the UTCs can compete. By funding university research and education, the USDOT is investing in our nation's intellectual transportation capacity.

Competitive University Transportation Center (UTC) Consortia	FY 2012 Request: [$80,000,000]

The University Transportation Centers (UTC) Program's mission is to advance transportation expertise and technology in the many disciplines that comprise transportation through education, research and technology transfer at university-based consortia.

The UTC Program provides a critical transportation knowledge base outside of the USDOT and addresses critical workforce needs for next generation of transportation leaders. DOT proposes to reform the UTC program by competitively selecting UTC consortia that will be governed by peer review principles.

All UTCs will be selected via rigorous competition that will include incentives for addressing key USDOT priorities (strategic goals, diversity, international collaboration, multi-state consortia, etc).

Performance metrics will ensure that transportation research and workforce needs are met, programmatic targets are realized, and that funds are being effectively invested. Reporting requirements will be strengthened to include explicit details of research results.

All UTCs will be diverse collaborations of more than one university. The UTC consortia will also advance transportation, education, and workforce development through degree-granting programs, seminars and training for practicing professionals, and outreach activities to attract new entrants to careers in transportation.

They will be multi-disciplinary, and multi-modal, with a focus on products that inform policy or spur innovation. UTCs will use the Transportation Research Board's Research in Progress and Transportation Research Information Service databases and will participate in USDOT Research Clusters via the www.transportationresearch.gov collaborative website to avoid duplication of efforts and to ensure wide awareness of efforts and dissemination of research results.

Each competitively selected UTC consortia will receive baseline funding. These competitively selected recipients will also be eligible to receive funds from the UTC Multimodal Competitive Research Grants.

RITA will receive UTC Program funding through an allocation from the Federal Highway Administration ($72,000,000) and a reimbursable agreement from the Federal Transit Administration ($8,000,000). The Secretary shall expend not more than 1.5 percent of amounts made available to carry out management and oversight of this program.

University Transportation Center (UTC) Multimodal FY 2012 Request: [$20,000,000]
Competitive Research

Most USDOT research is funded on a mode-by-mode basis. In order to encourage cross-modal research, the Secretary will appoint an internal cross-modal USDOT governance council to select annual priorities for targeted research needs. The University Transportation Center (UTC) Competitive Research Grant Program reserves 20 percent of the total $100M competitive UTC program to provide USDOT's modal administrations access to the nation's top academic researchers and University-based laboratories to address specific cross-modal research priorities in the areas of safety, state of good repair, economic competitiveness, livable communities and environmental sustainability. This program will allow all modal administrations to bring research needs, unanticipated issues, and quick-response problems to the table on an annual basis that UTC universities can compete for. This unique cross-modal program will enable USDOT staff to engage directly in partnership with UTC researchers to solve pressing problems or support policy decisions on a more nimble and responsive basis. Eligible recipients of these grants will be any of the universities participating in the competitively selected UTCs.

RITA will receive University Transportation Center (UTC) Multimodal Competitive Research Grants funding through an allocation from the Federal Highway Administration ($20,000,000). Frequently, UTCs develop promising results in their outlined baseline funded programs that modal administrations wish to capitalize on or to more specifically focus. These funds will allow modal administrations to develop Statements of Work to obtain that follow-on research to answer specific questions or to tap into unique expertise developed by a particular UTC consortium.

Multimodal Innovative Research Program   **TOTAL FY 2012 Request: [$20,000,000]**

The Multimodal Innovative Research Program is a restructuring of the current Advanced Research Program, managed by RITA. The existing Advanced Research Program is a reimbursable program with FHWA that funds research relevant to FHWA but also applicable to other modal administrations. The bulk of the projects in this program are managed by RITA as non-competitively awarded multi-year grants covering designated topic areas of research. One set of competitively awarded Cooperative Agreements covers a specific, congressionally specified area of research. Using the DOT RD&T Planning Team and RD&T Planning Council as cross-modal selecting bodies, the new program will fund a set of collaboratively outlined long-term research priorities. Competitively solicited proposals for the research will be open to bid by industry, university or state-based stakeholders and will serve as the basis for this Departmental, multi-modal research agenda. Guided by the USDOT RD&T Strategic Plan, this initiative will create opportunities for funding cross-modal research that is aimed at solving transportation problems at the interfaces between modes, or problems that affect more than one mode. The initiative will also include a component fostering creativity and innovation that can support a competition and prize program aimed at solving urgent transportation problems, and in support of the Administration's Open Government initiative. This program will also support key partnerships with other Federal agencies to fully leverage their investments in transportation research and product development to address transportation research issues.

The program will competitively award contracts for advanced multimodal transportation research to facilitate practical innovative approaches to solve transportation problems related to attainment of USDOT strategic goals and multi-modal elements of the DOT RD&T Strategic plan; to address issues affecting policy, and cross modal concerns such as efficient and intermodal goods and passenger movements; and to support development of advanced vehicle technologies and application/repurposing of existing technologies such as remote sensing & spatial information products.

Research products and results from this initiative will provide:
- Transportation system applications of advanced transportation technologies, methodologies, policies and decisions.
- Best practices in planning, operations, design and maintenance of transportation and related systems.
- Technology identification, modification and dissemination through outreach to other federal agencies, state and local transportation agencies and other public, private and academic stakeholders in the industry.

Successful projects will support U.S. DOT strategic goals by applying state-of-the art advanced technology solutions to multimodal transportation issues.

The program will focus on research to result in 'quick turnaround' products – applied vs. basic technology development including methodologies, policy guidelines, planning tools, prototypes/pilot products for practical application.

THIS PAGE HAS BEEN INTENTIONALLY LEFT BLANK

Transportation Safety Institute (TSI)
FY 2012 Funding: [20,000,000]

The Transportation Safety Institute in Oklahoma City, Oklahoma, is the Nation's leading provider of transportation safety training. The TSI educates more than 50,000 professionals each year in state-of-the-art safety methods and technologies.

The TSI assists DOT modal administrations in accomplishing their mission-essential training requirements. Since its inception, TSI has expanded its clientele to keep up with the needs of the Department and transportation industry. The Institute offers premier transit, aviation, motor carrier, traffic safety, hazardous material, and risk management training nationally and internationally.

The TSI supports several key strategies in the Department's Strategic Plan. It sponsors and participates in conferences, seminars, and meetings at which transportation consumers and providers can share advances in safety technology, regulation, and procedures. The TSI uses DOT web sites to communicate information on best safety practices, educational materials, consumer information, and other materials relating to safety. The TSI also improves safety in all modes through outreach, education, collaboration with public and industry safety partners, demonstration programs, consumer information and strategic media usage. Through training transportation industry safety professionals in accident investigation and prevention, TSI accomplishes these strategies and supports the Department's safety objectives. At the same time, TSI provides subject matter expertise to decision makers and the public in transportation safety.

How TSI Operates
- Direction and budget oversight provided by RITA
- Funded via reimbursable agreements, tuitions, and fees
- Associate and contract staff (industry experts) are instrumental in delivery/development of the training.

THIS PAGE HAS BEEN INTENTIONALLY LEFT BLANK

RESEARCH AND INNOVATIVE TECHNOLOGY ADMINISTRATION
VOLPE NATIONAL TRANSPORTATION SYSTEMS CENTER
Working Capital Fund
Program and Performance

The Working Capital Fund finances multidisciplinary research, evaluation, analytical and related activities undertaken at the Volpe Center in Cambridge, MA. The fund is financed through negotiated agreements with the Office of the Secretary, Departmental operating administrations, and other governmental elements requiring the Center's capabilities. These agreements also define the activities undertaken at the Volpe Center.

Volpe National Transportation Systems Center
Working Capital Fund

The Working Capital Fund finances multidisciplinary research, evaluation, and analytical and related activities under the John A. Volpe National Transportation Systems Center in Cambridge, MA. The Volpe Center is solely financed through negotiated agreements with the Office of the Secretary, U.S. DOT modal administrations, and other governmental elements requiring the Volpe Center's capabilities.

The Volpe Center mission is to advance technical, operational, institutional and managerial innovations to improve the nation's transportation system; to anticipate emerging transportation issues; and to be a federal center of excellence for informed decision-making. The Volpe Center provides a unique capability in the synergy of transportation expertise with its institutional knowledge of the global transportation system and its stakeholder perspective; the federal perspective with its awareness of federal responsibilities, objectives and activities in the public interest; and experience with the full spectrum of technologies and disciplines relevant to transportation system improvements. Five hundred fifty federal employees, all dedicated transportation professionals - half of which have advanced degrees, is the Volpe Center's most important asset who represent a full spectrum of disciplines from engineering to physical and social sciences.

To facilitate its mission and the mission of the Department, the Volpe Center has eight Centers of Innovation: Multimodal Systems Research and Analysis; Safety Management Systems; Environmental and Energy Systems; Freight Logistics and Transportation Systems; Physical Infrastructure Systems; Communication, Navigation, Surveillance and Traffic Management Systems; Human Factors Research and System Applications; and Advanced Vehicle and Information Network Systems. Collectively these Centers of Innovation provide progressive, multidisciplinary expertise in support of key U.S. DOT and national transportation issues and in solving complex transportation problems.

The Volpe Center also provides leadership to advise, facilitate and manage a wide range of high profile, complex, often rapid response, U.S. DOT, agency, and priority initiatives of the Secretary, RITA Administrator and Volpe Center Director. Through extensive outreach and coordination the Volpe Center supports the U. S. DOT's efforts to expand its understanding of current and emerging issues through strategic planning initiatives and timely communication.

DEPARTMENT OF TRANSPORTATION
RESEARCH AND INNOVATIVE TECHNOLOGY ADMINISTRATION
WORKING CAPITAL FUND,
VOLPE NATIONAL TRANSPORTATION SYSTEMS CENTER
PROGRAM AND FINANCING
(in thousands of dollars)

Identification code 69-4522-0-4-407

		FY 2010 ACTUAL	FY 2011 CR (ANNUALIZED)	FY 2012 REQUEST
	Obligations by program activity:			
0801	Volpe National Transportation Systems Center	250,893	250,000	250,000
0900	Total new obligations	250,893	250,000	250,000
	Budgetary Resources:			
	Unobligated balance:			
1000	Unobligated balance brought forward, Oct 1	236,534	260,963	260,963
1050	Unobligated balance (total)	236,534	260,963	260,963
	Budget authority:			
	Spending authority from offsetting collections, discretionary:			
1700	Collected	277,528	250,000	250,000
1701	Change in uncollected payments, Fed sources	(2,206)	-	-
1750	Spending authority from offsetting collections, disc (total)	275,322	250,000	250,000
1930	Total budgetary resources available	511,856	510,963	510,963
	Memorandum (non-add) entries:			
1941	Unexpired unobligated balance, end of year	260,963	260,963	260,963
	Change in Obligated Balance:			
	Obligated balance, start of year (net):			
3000	Unpaid obligations, brought forward, Oct 1 (gross)	117,659	114,461	114,461
3010	Uncollected payments, Fed sources, brought forward, Oct	(125,859)	(123,650)	(123,650)
3020	Unobligated balance, start of year (net)	(8,198)	(9,189)	(9,189)
3030	Obligations incurred, unexpired accounts	250,893	250,000	250,000
3040	Outlays (gross)	(254,091)	(250,000)	(250,000)
3050	Change in uncollected payments, fed sources, unexpired	2,206		
	Obligated balance, end of year (net):			
3090	Unpaid obligations, end of year (gross)	114,461	114,461	114,461
3091	Uncollected payments, fed sources, end of year	(123,650)	(123,650)	(123,650)
3100	Obligated balance, end of year (net)	(9,189)	(9,189)	(9,189)
	Budget Authority and Outlays, net:			
	Discretionary:			
4000	Budget authority, (gross)	275,322	250,000	250,000
4010	Outlays from new discretionary authority	174,583	250,000	250,000
4011	Outlays from discretionary balances	79,508	-	-
4020	Outlays, gross (total)	254,091	250,000	250,000
	Offsets against gross budget authority and outlays:			
	Offsetting collections (collected) from:			
4030	Federal sources	(274,956)	(250,000)	(250,000)
4033	Non-federal sources	(2,573)	-	-
4040	Offsets against gross budget authority and outlays (total)	(277,529)	(250,000)	(250,000)
	Additional offsets against gross budget authority only:			
4050	Change in uncollected payments, Fed sources, unexpired	2,206	-	-
4060	Additional offsets against budget authority only (total)	2,206	-	-
4070	Budget authority, net (discretionary)	-	-	-
4080	Outlays, net (discretionary)	(23,438)	-	-
4180	Budget authority, net (total)	-	-	-
4190	Outlays, net (total)	(23,438)	-	-

DEPARTMENT OF TRANSPORTATION
RESEARCH AND INNOVATIVE TECHNOLOGY ADMINISTRATION

WORKING CAPITAL FUND
VOLPE NATIONAL TRANSPORTATION SYSTEMS CENTER

Object Classification (in thousands of dollars)

Identification code 69-4522-0-4-407

Reimbursable Obligations

Obj Code	Item	FY 2010 ACTUAL	FY 2011 CR (ANNUALIZED)	FY 2012 REQUEST
Personnel compensation				
2 111	Full-time permanent	50,840	50,000	50,000
2 113	Other than full-time permanent	3,429	3,000	3,000
2 115	Other personnel compensation	1,301	1,000	1,000
2 119	Total personnel compensation	55,570	54,000	54,000
2 121	Civilian personnel benefits	15,160	14,000	14,000
2 130	Benefits for former personnel	26	0	0
2 210	Travel & transportation of persons	3,708	5,000	5,000
2 220	Transportation of things	143	0	0
2 233	Commun, utilities & misc. charges	2,927	4,000	4,000
2 240	Printing and reproduction	170	0	0
2 251	Advisory and assistance services	2,839	0	0
2 252	Other services	61,400	60,000	60,000
2 253	Purch of G&S from Govt accounts	72	1,000	1,000
2 254	O&M of facilities	4,423	5,000	5,000
2 255	R&D Contracts	86,832	94,000	94,000
2 257	O&M of equipment	415	1,000	1,000
2 260	Supplies and materials	4,822	1,000	1,000
2 310	Equipment	11,036	8,000	8,000
2 320	Land and structures	1,352	3,000	3,000
2 990	Subtotal, reimbursable obligations	250,895	250,000	250,000
9 999	Total new obligations	250,895	250,000	250,000

EMPLOYMENT SUMMARY

		FY 2010 ACTUAL	FY 2011 CR (ANNUALIZED)	FY 2012 REQUEST
	Reimbursable:			
2 001	Civilian full-time equivalent employment	532	550	550

Detailed Justification for the Intelligent Transportation Systems (ITS): Core Program

What Do I Need to Know Before Reading This Justification?

The funding request of $110 million remains unchanged from prior years.

What Is the Request and What Will We Get for The Funds?

FY 2012 Intelligent Transportation Systems (ITS):
Core Program
Budget Request
($000)

Program Activity	FY 2010 Actual	FY 2012 Request	Change FY 2010-2012
Research	[80,868]	[90,600]	[9,732]
Technology Transfer and Evaluation	[14,702]	[13,900]	[-802]
ITS Program Support	[7,280]	[5,500]	[-1,780]
Total	[$102,850]	[$110,000]	[$7,150]

Key actions:
- Develop and test proven reliable multimodal vehicle-to-vehicle and vehicle-to-infrastructure communications safety applications using dedicated short range communications to demonstrate whether a positive cost benefit analysis will support a future regulatory decision to require this technology in all new vehicles.
- Develop, test and support deployment of smart multimodal infrastructure applications for safety, mobility and environmental sustainability purposes to allow state DOTs and other local transportation agencies to deploy and operate smart infrastructure.
- Establish one or more a multi-source data environments to enable transformative operations efficiency and environmental sustainability for transportation managers of freeways, arterials, transit systems, maritime, and rail systems.
- Develop and test applications that can directly measure fuel consumption and emissions using vehicle-to-vehicle wireless technology to improve environmental sustainability.

What is This Program?

Strategic Goals: The Intelligent Transportation Systems program specifically supports five of DOT's strategic goals: Safety, Environmental Sustainability, Livable Communities, Economic Competitiveness, and Organizational Excellence.

Purpose and Beneficiaries: To fulfill the research role previously outlined by Congress and in support of the USDOT strategic goals. The Intelligent Transportation Systems program consists of a multi-modal, mode-specific, and exploratory research administered as a collaborative partnership between the Intelligent Transportation Systems program and USDOT modal administrations. Beneficiaries include the traveling public, state DOTs, the automotive manufacturers, transit agencies, other local transportation agencies, private industry data providers, application developers, academia, and international transport agencies.

Partners: The Intelligent Transportation Systems program jointly conducts research with our modal partners, including:
- Federal Highway Administration (FHWA)
- Federal Motor Carrier Safety Administration (FMCSA)
- Federal Railroad Administration (FRA)
- Federal Transit Administration (FTA)
- Maritime Administration (MARAD)
- National Highway Traffic Safety Administration (NHTSA)

Prior Year Accomplishments:
The Intelligent Transportation Systems program will conduct a large scale safety pilot of the vehicle-to-vehicle and vehicle-to-infrastructure communications safety technologies in approximately 2,000–3,000 test vehicles in multiple geographic locations around the country to demonstrate and evaluate the effectiveness of the safety applications. Real time data capture and management testing will be completed to validate data quality, reliability, and consistency to ensure it supports operational requirements of transportation managers and information requirements for users.

Vehicle-to-vehicle and vehicle-to-infrastructure communications technical standards will be approved, and the vehicle-to-vehicle and vehicle-to-infrastructure communications test bed in Detroit, Michigan, will be upgraded to support the Safety Pilot testing. Real time data capture and management testing will be conducted to inform researchers on the best way to collect, control, and utilize real time data for performance management purposes. International standards harmonization, support of deployment through technology transfer and outreach and broadening the range of stakeholders involved in the program will also occur.

Why Is This Program Necessary?

Intelligent Transportation Systems technology solutions offer the opportunity to reduce crashes, fatalities and injuries thereby improving public health; improve efficiency, reliability and mobility; and reduce fuel consumption and greenhouse gas emissions. The travelling public, automotive manufacturers, and state and local transportation agencies will all benefit from reduced crashes, injuries, and fatalities on the U.S. transportation system. Vehicle-to-vehicle and vehicle-to-infrastructure communications safety applications are designed to increase situational awareness and reduce or eliminate crashes through data transmission that supports: driver advisories, driver warnings, and vehicle and/or infrastructure controls. With these multimodal applications for light vehicles, trucks, buses, and fleets of all kinds, Vehicle-to-vehicle and vehicle-to-infrastructure communications may potentially address approximately 80 percent of crash scenarios with unimpaired drivers, preventing tens of thousands of automobile crashes every year (further research will incorporate heavy vehicle crashes including buses, motor carriers, and rail).

The traveling public, freight movers, and state and local transportation agencies will benefit from the use of real time data applications to improve the efficiency, mobility, reliability and sustainability of the U.S. transportation system. Data generated from vehicle-to-vehicle and vehicle-to-infrastructure communications systems can provide travelers and operators with detailed, real-time information on vehicle location, speed, emissions and other operating conditions. Thus informed travelers and commercial freight movers can use this information to improve system operations and choose optimum routing to avoid congested routes, take alternate routes, use other modes of transportation, shift to public transit, or reschedule their trips — all of which can save time, fuel, and money.

Without this funding, the adoption and deployment of Intelligent Transportation Systems by state and local transportation agencies would be reduced. Without federal research investments, state DOTs, transit properties, and transportation managers would not have the resources to conduct independent Intelligent Transportation Systems research and implement successful research results.

Private sector innovations would be developed for technologies that have market demand or a clear business case. However, due to the high risk and cost involved with safety applications, there is not a clear business case for their development of vehicle-to-vehicle or vehicle-to-infrastructure without federal research and government sponsored standards development to ensure interoperability.

How Do You Know the Program Works?

Effective: The amount of Intelligent Transportation Systems technology deployment by state DOT's and transit agencies has been increasing over time and our transportation system has benefitted from this deployment in terms of increased mobility, better emergency and incident response times, better overall traffic management, and increased capacity during rush hour. Additionally, private sector companies are also investing dollars into vehicle-to-vehicle research with the expectation that the successful testing and regulatory decision by NHTSA will require vehicle-to-vehicle technology in U.S. vehicles.

Research: Positive results from our vehicle-to-vehicle proof of concept testing in Michigan have demonstrated that the vehicle-to-vehicle and vehicle-to-infrastructure approach is feasible and effective. Independent evaluations are conducted on all research projects and made available to the public through the USDOT library. Effective strategies and technologies are adopted such as some of the congestion initiative and rural safety applications that have been demonstrated in specific operational locations around the country.

Sub-program Balance: The Intelligent Transportation Systems Research sub-program is critical to achieving the Departmental goals. However, the Intelligent Transportation Systems program could not be properly executed and deployment could not be made possible without the Program Support and Technology Transfer and Evaluation subprograms.

Efficient: Vehicle-to-vehicle research is currently being conducted to enable a regulatory decision by the USDOT in the 2013 timeframe. The efficient planning, acquisition, demonstration and evaluation of a complex research workload are supporting this aggressive regulatory timeframe.

Why Do We Want/Need To Fund The Program At The Requested Level?

Funding Options: The current funding level is adequate to execute the scope of research outlined in the Intelligent Transportation Systems Strategic Research Plan. The Intelligent Transportation Systems program has been funded at $110 million for the entire 18 years of the program's existence. Thus, the program has matured and is programmatically operating at that level of funding. If funding were increased or decreased, the supporting procurement activities and program support would increase or decrease accordingly. However, achieving the vehicle-to-vehicle regulatory decision requires a large scale research and testing program with sufficient data to enable NHTSA to make a positive decision. If funding levels were decreased, this program could not be executed as described in the Intelligent Transportation Systems Strategic Research Plan.

History/Out Year Needs: The Intelligent Transportation Systems Strategic Research Plan has identified a multi-year scope of research based on a $110 million funded program. The out year needs are consistent with the requested funding amount.

Detailed Justification for the Intelligent Transportation Systems (ITS) Program: Wireless Innovation and Infrastructure Initiative

What Do I Need To Know Before Reading This Justification?

This FY 2012 request is for a new $100 million initiative in conjunction with the President's Wireless Innovation and Infrastructure Initiative, Wireless Innovation (WIN) Fund. A separate legislative package will transmit this mandatory request of $100 million. This is in addition to the $110 million Intelligent Transportation Systems (ITS) core research and technology transfer program. The separate ITS core program funding request of $110 million remains unchanged from prior years.

What Is The Request And What Will We Get For The Funds?

Request: $100 million to be used over a five year period, to establish a new multi-modal wireless technology initiative to create a safer, less congested, more flexible, efficient and resilient transportation system.

Actions: The proposed wireless research, technology, testing and model deployment program will develop "living laboratories" while leveraging the existing transportation infrastructure, where innovative wireless communications methods and applications can be developed safely to advance into deployment.

FY 2012 Intelligent Transportation Systems (ITS): Wireless Innovation and Infrastructure Initiative Budget Request ($000)

Program Activity	FY 2010 Actual	FY 2012 Request	Change FY 2010-2012
Wireless Innovation and Infrastructure Initiative	[0]	[100,000]	[100,000]
Total	[$0]	[$100,000]	[$100,000]

Key actions:
- Demonstrate "living laboratories" in a competitively-selected region/corridor that leverage other public private investments for safety, mobility, emergency response, energy and environmental benefits of real-time wireless applications, including passenger and freight benefits in autos, trucks, rail, transit, ports and inland waterways, and at border crossings.
- Create wireless "fast lanes" for multi-modal applications for real time inspections, reporting, and access nationwide, including underserved rural areas and border crossings.
- Work with state inspection and public safety partners, along with other Federal agencies to deploy rural wireless access points in areas of critical need for enhanced emergency communications.
- Require that all applications discourage distracted driving/operations.

Key Outputs: Additional details of this proposal will be presented in the legislative package that will contain this $100 million mandatory request. However, possible deployments could yield the following: enhanced cybersecurity of vehicle communications; location of alternative energy stations, electric vehicle charging stations, and other wireless power transfer units (e.g., auxiliary power units, port power); real-time, wireless traffic management centers with wireless information sharing with emergency operations centers; real-time traffic management strategies driven by vehicle-infrastructure wireless communication at local access points; multi-modal route planning, optimization and notifications (e.g. road, rail, transit, commercial vehicle, freight flow, fleets, etc); real-time information and safety enhancements for drivers, commercial vehicles, transit passengers, pedestrians, motorcyclists, and bicyclists; safety compliance checks of commercial motor vehicles at highway speeds; accessibility for disabled, elderly, or disadvantaged individuals; and universal parking and transit payment applications.

Key Outcomes: Increased safety, efficient, capacity, and in all modes of surface transportation; improved energy use and environmental performance from the transportation system; improved livability through better transit, disability access, emergency response and bicycle/pedestrian use, reduced emissions, and decreased congestion.

What Is The Program?

Strategic goals: Safety, Environmental Sustainability, Livable Communities, and Economic Competiveness.

Purpose/beneficiaries: This program will develop "living laboratories" in the existing infrastructure, where innovative wireless communications methods and applications can be developed safely *in-situ* and may be rapidly advanced into deployment.

Partners: DOT surface modes: FHWA, NHTSA, FMCSA, FTA, FRA, MARAD; other Federal departments (e.g. DoD, DOE, EPA, DHS etc.); public and private safety advocates; academia; university transportation centers; and industry.

Description: This additional $100 million from the WIN fund will be used over a five year period, and will provide the Intelligent Transportation Systems (ITS) program and its stakeholders the ability to seek new and innovative opportunities to pursue ground-breaking research and competitive deployments of wireless technology applications.

Base budget/prior year accomplishments: This is a new initiative.

Why is This Particular Program Necessary?

Benefits: Deployment of the resulting proven ITS applications will greatly advance the safety, efficiency, convenience, and environmental sustainability of surface transportation.

Consequences of lack of funding: Missed opportunities to pursue ground-breaking research and competitive deployments of wireless technology applications.

How Do You Know the Program Works?

Effective: Demand for safe, reliable and resilient communications, cybersecurity, and computing infrastructures from transportation advocates, and technology developers.

Research: DOT proposes to work with innovators to address the need for market development in wireless-based transportation applications. Potential research includes competitively selecting one or more regions or travel/freight corridors as an integrated, multi-modal national testbed for wireless applications; identification of locations across the U.S. where these applications could be demonstrated; secure, reliable and resilient communications; applications that discourage distracted driving/operations; wireless "fast lanes" for vehicle inspections, wireless research, technology, testing and model deployment program. Requiring that all applications discourage distracted driving/operations.

Sub-program balance: N/A

Efficient: The ITS JPO, RITA, is the designated federal agency for the pursuit of intelligent transportation systems research and technology transfer and leads multi-modal ITS research across the DOT surface modes.

Why Do We Want/Need To Fund The Program At The Requested Level?

Funding options: $100 million over a five year period.

History/outyear needs: None

THIS PAGE HAS BEEN INTENTIONALLY LEFT BLANK

EXHIBIT III-1

BUREAU OF TRANSPORTATION STATISTICS
SUMMARY BY PROGRAM ACTIVITY
Appropriations, Obligation Limitations, and Exempt Obligations
($000)

	FY 2010 ENACTED	FY 2011 CR (ANNUALIZED)	FY 2012 REQUEST	Change FY 2011-2012
Bureau of Transportation Statistics				
Safety Data and Analysis	-	-	3,049	3,049
Travel Statistics	2,947	2,947	3,064	117
Freight Statistics	10,723	10,723	16,021	5,298
Transportation Economics	1,811	1,811	1,938	127
Geospatial Information Systems	1,758	1,758	1,521	-237
Transportation Analysis, Data Quality and Performance Metrics	7,416	7,416	7,268	-148
National Transportation Library	2,345	2,345	2,139	-206
TOTAL: [Discretionary] [1/]	[27,000]	[27,000]	[35,000]	[8,000]
Direct FTE	68	70	70	0
Reimbursable FTE	13	19	19	0

[1/] Resources are shown as non-adds because the Bureau of Transportation Statistics is an allocation account under the Federal-Aid Highways program.

BUREAU OF TRANSPORTATION STATISTICS
SUMMARY BY PROGRAM AND ADMINISTRATIVE EXPENSES

ACCOUNT NAME	FY 2011 CR (Annualized)			FY 2012 President Request			Change Admin FY 2011-2012	Change Program FY 2011-2012	Change Total FY 2011-2012
Bureau of Transportation Statistics	Admin	Program	Total	Admin	Program	Total			
Safety Data and Analysis	-	-	-	1,049	2,000	3,049	1,049	2,000	3,049
Travel Statistics	2,497	450	2,947	2,614	450	3,064	117	0	117
Freight Statistics	6,223	4,500	10,723	5,606	10,415	16,021	-617	5,915	5,298
Transportation Economics	1,651	160	1,811	1,778	160	1,938	127	0	127
Geospatial Information Systems	1,398	360	1,758	1,161	360	1,521	-237	0	-237
Transportation Analysis, Data Quality and Performance Metrics	7,021	395	7,416	6,873	395	7,268	-148	0	-148
National Transportation Library	1,997	348	2,345	1,791	348	2,139	-206	0	-206
TOTAL:	20,787	6,213	27,000	20,872	14,128	35,000	85	7,915	8,000

EXHIBIT III-1a

BUREAU OF TRANSPORTATION STATISTICS
SUMMARY ANALYSIS OF CHANGE FROM FY 2011 TO FY 2012
Appropriations, Obligation Limitations, and Exempt Obligations

($000)

Item	Change from FY 2011 to FY 2012	FTE Change from FY 2011 to FY 2012
FY 2011 CR (ANNUALIZED)	27,000	70
BUREAU OF TRANSPORTATION STATISTICS		
Adjustments to Base		
One Less Compensable Day	-47	
Working Capital Fund Inc/Dec	107	
Inflation/Deflation	25	
Subtotal, Adjustments to Base	85	0
New or Expanded Programs		
Safety Data and Analysis	2,000	
Travel Statistics	0	
Freight Statistics	5,915	
Transportation Economics	0	
Geospatial Information Sysems	0	
Transportation Analysis, Data Quality and Performance Metrics	0	
National Transportation Library	0	
Subtotal, New or Expanded Program Increases/Decreases	7,915	
Total FY 2012 Request	35,000	70

EXHIBIT III-2

RESEARCH AND INNOVATIVE TECHNOLOGY ADMINISTRATION
ANNUAL PERFORMANCE RESULTS AND TARGETS

The Research and Innovative Technology Administration (RITA) - Bureau of Transportation Statistics (BTS) integrates performance results into its budget request to demonstrate alignment with the Department of Transportation's Strategic Plan. RITA/BTS tracks the following agency measures in support of the USDOT strategic goals of Safety, Economic Competitiveness, Livability, Environmental Sustainability, and Organizational Excellence.

USDOT Goal/Outcome: Organizational Excellence

Total Congressional Offices, Federal Agencies, and State USDOT Offices contacting BTS for information	2007	2008	2009	2010	2011	2012
Target		---	---	Baseline		
Actual		---	---	1,686*		

*This includes 852 questions to the NTL ARRA Public Response Team, which will sunset in 2010.

USDOT Goal/Outcome: Safety

Number of Times Safety Data is Accessed	2007	2008	2009	2010	2011	2012
Target		---	---	---	---	---	Baseline
Actual		---	---	---	---	---	

USDOT Goal/Outcome: Economic Competitiveness

Number of Times Travel Related Data is Accessed	2007	2008	2009	2010	2011	2012
Target		---	---	---	Baseline	TBD	TBD
Actual		---	---	---			

USDOT Goal/Outcome: Economic Competitiveness

Number of Times Freight Data is Accessed	2007	2008	2009	2010	2011	2012
Target		---	---	---	Baseline	TBD	TBD
Actual		---	---	---			

USDOT Goal/Outcome: Economic Competitiveness

Number of Times Transportation Economics Data (Transportation Services Index) is Accessed	2007	2008	2009	2010	2011	2012
Target	---	---	---	Baseline	TBD	TBD
Actual	---	---	---			

USDOT Goal/Outcome: Livable Communities/Safety/Environmental Sustainability

Number of Times National Transportation Atlas Database is Accessed	2007	2008	2009	2010	2011	2012
Target	---	---	---	Baseline	TBD	TBD
Actual	---	---	---			

USDOT Goal/Outcome: Economic Competitiveness

Number of Times Transportation Analysis Data (National Transportation Statistics, Key Transportation Indicators, Transportation Statistics Annual Report)	2007	2008	2009	2010	2011	2012
Target	---	---	---	Baseline	TBD	TBD
Actual	---	---	---			

USDOT Goal/Outcome: Economic Competitiveness

Average Daily Visits to National Transportation Library	2007	2008	2009	2010	2011	2012
Target	---	---	---	Baseline	TBD	TBD
Actual	---	---	---			

USDOT Goal/Outcome: Economic Competitiveness

Number of Times National Transportation Library Electronic Database is Accessed	2007	2008	2009	2010	2011	2012
Target	---	---	---	Baseline	TBD	TBD
Actual	---	---	---			

USDOT Goal/Outcome: Economic Competitiveness

Number of Times Airline Data is Accessed	2007	2008	2009	2010	2011	2012
Target	---	---	---	Baseline	TBD	TBD
Actual	---	---	---			

RESEARCH AND INNOVATIVE TECHNOLOGY ADMINISTRATION
BUREAU OF TRANSPORTATION STATISTICS
(Allocation Account under FHWA's Federal-Aid Highway)
OBJECT CLASSIFICATION
(In thousands of dollars)

		FY 2010 ACTUAL	FY 2011 (ANNUALIZED)	FY 2012 REQUEST
Direct Obligations:				
	Personnel compensation:			
1111	Full-time permanent	6,270	9,521	9,484
1112	Other than full-time permanent	5	-	-
1113	Other personnel compensation	130	162	161
1115	Other personnel compensation	160	204	204
	Total Personnel Compensation	6,565	9,887	9,849
1121	Civilian Personnel benefits	1,601	2,314	2,305
1210	Travel and Transportation of persons	127	205	206
1220	Transportation of Things	0	-	-
1231	Rent to GSA	1,437	1,443	1,450
1240	Printing & Production	70	10	10
1251	Advisory and Assistance services	4,981	7,659	7,303
1252	Other Services	223	343	327
	Other purchases of goods and services			
1253	from gov't accounts	7,174	11,028	10,516
1257	Operation and maint of equipment	1,527	2,349	2,240
1260	Office Supplies	33	40	40
2310	Equipment	650	750	754
1990	Subtotal, direct obligations	24,388	36,028	35,000
1990	Reimbursable obligations	8,364	10,000	10,000
1990	Total obligations	32,752	46,028	45,000

Personnel Summary:

		FY 2010 ACTUAL	FY 2011 (ANNUALIZED)	FY 2012 REQUEST
	Reimbursable:			
2001	Civilian Full-time Equivalent Employment	13	19	19
	Direct:			
3001	Civilian Full-time Equivalent Employment	68	70	70

RESEARCH AND INNOVATIVE TECHNOLOGY ADMINISTRATION
BUREAU OF TRANSPORTATION STATISTICS
10-Year Funding History
(In thousands of dollars)

YEAR	REQUEST	ENACTED	
2003	35,806	30,499	[1]
2004	35,239	30,235	[2]
2005	32,199	30,015	[3]
2006	32,869	26,730	[4]
2007	27,480	27,562	[5]
2008	27,000	27,000	[6]
2009	27,000	27,000	[7]
2010	28,000	27,000	[8]
2011	30,000	-	
2012	35,000	-	

[1] FY 2003 reflects a reduction of $300,000 for WCF expenses (section 362) and .65% rescission of $201,500 (section 601) of P.L. 108-7.

[2] FY 2004 reflects a reduction of $581,000 for WCF expenses (section 517) and .59% across the board rescission of $183,000 (section 168) of P.L. 180-199.

[3] FY 2005 reflects a reduction of $737,000 to WCF expenses (section 197) as stated in P.L. 108-477.

[4] FY 2006 reflects a 1% across the board rescission of $270,000 as stated in P.L. 109-148, section 3801.

[5] FY 2007 reflects levels under a year long CR. An increase of $562,000 over amount is due to Revenue Aligned Budget Authority (RABA) estimates ($462,000) and a pay increase ($93,000) provided by H.J. Res 20.

[6] FY 2008 reflects funding provided in P.L. 110-161.

[7] FY 2009 reflects funding provided in P.L. 111-8.

[8] FY 2010 reflects funding provided in P.L. 111-117.

Detailed Justification for BTS Safety Data and Analysis Program (SDAP)

What Do I Need To Know Before Reading This Justification?

The Safety Data and Analysis Program is designed to centralize, standardize and address gaps in safety data across all modes in support of the USDOT Safety Council. The safety data will allow the Safety Council to develop a formal process for multi-modal data sharing and adopt a data-driven approach to identify, analyze, evaluate, and potentially predict systemic problems and create improvements across modes and sectors.

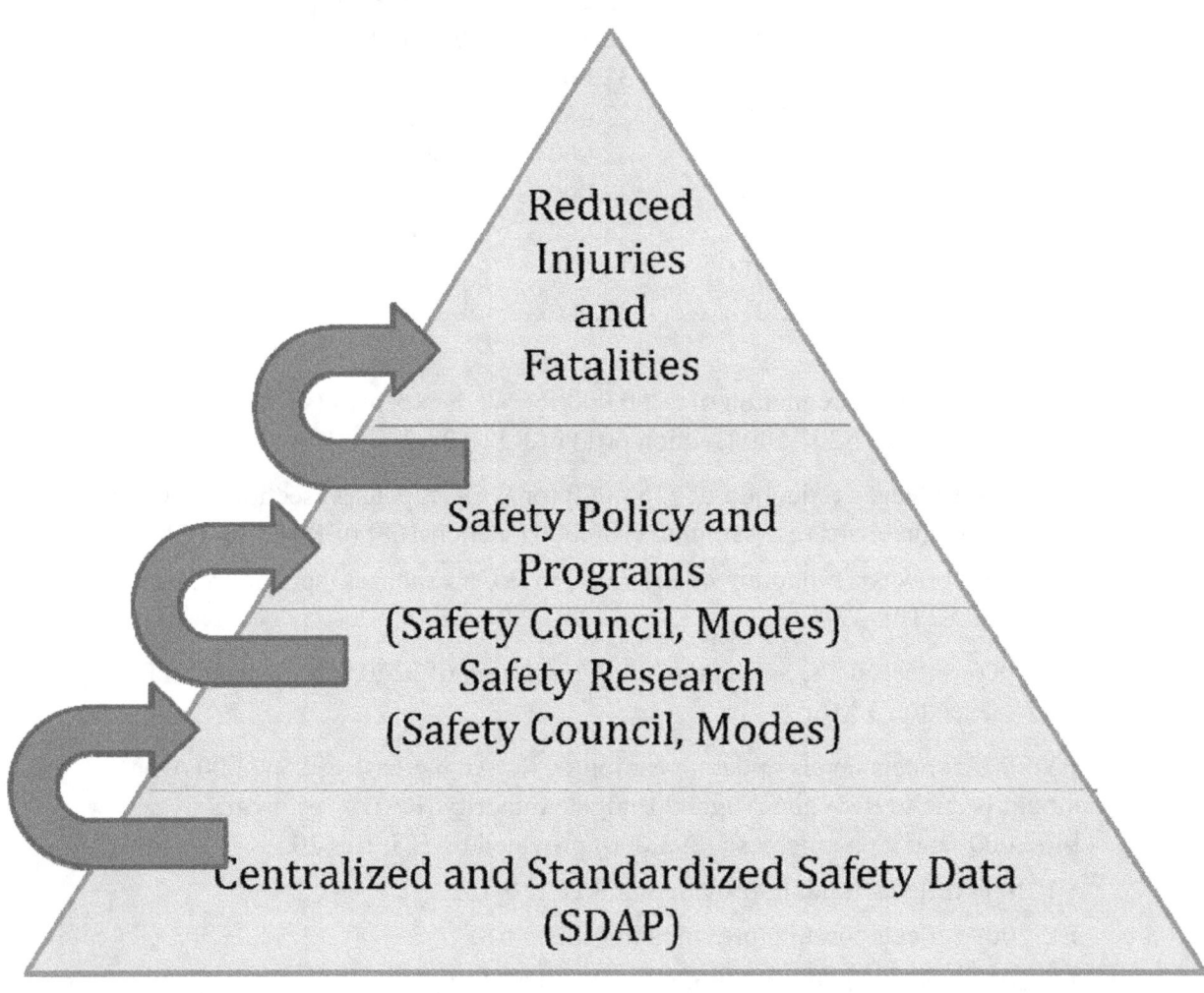

- This is a new collaborative initiative to address the synthesis, collection, processing and analysis of multimodal transportation safety data.
- Safety is USDOT's highest priority.

What Is The Request And What Will We Get For The Funds?

FY 2012 Transportation Safety Data Program
Budget Request
($000)

Program Activity	FY 2010 Actual	FY 2012 Request	Change FY 2010-2012
Safety Data Program	0	3,049	3,049
TOTAL[1]	**0**	**$3,049**	**$3,049**

[1] The total funding includes salaries and administrative expenses and contract program.

Key Actions:
- Coordinate with other modal administrations to create a collaborative portal, available to a broad range of stakeholders and the public, for accessing, synthesizing and analyzing safety data across the USDOT, including precursor (near-miss) and incident/accident safety information;
- Integrate multimodal Intelligent Transportation System (ITS) deployment data with safety data;
- Coordinate and expand the collection of multimodal transportation exposure data to improve estimation of safety risk rates and facilitate risk analysis across modes (with a focus on special sub-populations, such as motorcyclists, pedestrians and bicyclists);
- Develop and apply new measures of safety risk exposure in coordination with the other modal administrations to better characterize and communicate risk;
- Coordinate collection and analysis of occupational safety and health data pertinent to the analysis of vehicle operator and safety performance, including actual video of drivers under various real life conditions;
- Develop advanced statistical methodology for emerging areas of study, such as naturalistic driving studies in USDOT;
- Provide technical support to the modal administrations in the design and implementation of surveillance data systems, data standardization, and integration of safety data with GIS standards;
- Provide support to the modal administrations in the areas of study design, study protocol development, data collection and analysis of transportation safety studies.

Key Outputs:
- Development of data clearinghouse to support the analysis of multimodal USDOT safety data; inventory and synthesis of safety data, new safety risk exposure measures across the modal administrations. The clearinghouse will have a group of expert staff to support the program.
- Technical support to improve data collection and analysis; document critical multimodal data gaps for USDOT, state DOTs, the academic and research communities, and other transportation stakeholders.

Key Outcomes:
- Improved safety program assessments and to support more efficient USDOT safety investments.
- A focused effort to reducing injuries and the approximately 34,000 annual transportation-related fatalities by improving access to safety data, including information on near-miss (precursor) data.

What Is This Program?

The Safety Data and Analysis Program (SDAP) is a new initiative in FY 2012 to enable BTS to effectively integrate safety data across modes and address gaps in existing USDOT data programs. Improving safety throughout the transportation sector is USDOT's highest priority. To achieve this goal, Secretary LaHood has created a USDOT Safety Council to work collaboratively to address safety problems by leveraging resources, sharing experience, expertise and best practices. The Safety Council's strategy is to adopt a data-driven approach to identify, analyze, evaluate, and potentially predict systemic problems and create improvements across modes and sectors.

Partners: FHWA, NHTSA, FMCSA, FRA, MARAD, PHMSA, FTA, FAA, USDOT Safety Council, safety advocates and transportation industry.

Description: The SDAP addresses multimodal data issues and the DOT strategic goal of improving safety through accurate and timely potential incident and exposure data and analyses. This program augments USDOT's existing data collection and data analysis efforts, providing a portal for integrated data for research and decision making by critical stakeholders.

Why Is This Particular Program Necessary?

The SDAP will provide improved multimodal safety data and analysis through increased collaboration across modal administrations. The centralized data portal will provide standardized access to a wide array of data sources, and will identify and address gaps in USDOT data programs and identify high concentrations of transportation safety risk. Transportation risk is highly concentrated in certain sub-populations and occurs under certain conditions; e.g., in aviation on takeoffs and landings, especially at night; and in motor vehicle travel, especially with higher risk drivers.

The transportation safety data collected, processed, maintained, and disseminated by the different USDOT modal administrations lack uniformity and comparability in terminology and reporting formats (e.g., different injury reporting criteria), which may limit system-wide risk assessments and multimodal data analysis. In addition, there are gaps in existing data, such as a lack of complete information on operator fatigue and its impact on safety performance.

Without the requested funding, BTS will not be able to provide a multi-modal data portal and set of cross-cutting analysis capabilities in support of the USDOT's highest priority: safety. There would continue to be a significant number of unfilled gaps in safety data across the modes, and safety data would remain scattered among the modal administrations as opposed to being accessible through a centralized portal.

How Do You Know the Program Works?

The likelihood of success for the requested program will be indicated by several factors:

Effectiveness: As an example, the preliminary results from the current confidential close-calls reporting system for rail has indicated that close-calls reporting systems are needed across modes to support safety initiatives.

Research: An NTSB report has recommended that BTS develop better risk exposure data: "in reviewing BTS efforts to establish data quality standards, identify information gaps, and ensure compatibility between DOT safety data systems, the Safety Board recognizes a number of important BTS accomplishments. BTS has led safety data improvement efforts in recent years, and the Safety Board commends the DOT's efforts in this area."

Efficient: BTS, as a designated federal statistical agency, has authority to collect and protect confidential information. BTS is organized to collect and analyze cross-modal safety data.

Why Do We Want/Need To Fund The Program At The Requested Level?

The requested level of funding is needed to:
- Develop specifications for a safety data portal to serve DOT and other transportation safety communities;
- Develop data harmonization processes for cross-walking of various modal data sources to facilitate development of exposure measures;
- Establish and maintain analytic expertise for safety risk measurement and analysis, collection and analysis of naturalistic driving studies, and potential development and operation of close-calls reporting systems for other modes.

Detailed Justification for Passenger Travel Statistics Program

What Do I Need To Know Before Reading This Justification?

The Passenger Travel Statistics Program is the primary way the USDOT collects data on the way Americans travel around their communities and around the country. This information is integral for the USDOT to base its very significant investment and policy decisions on sound science and rigorous analysis.

The Passenger Travel Statistics Program provides critical information for performance measures associated with the USDOT's strategic goals for safety, livable communities, state of good repair, and environmental sustainability.

What is The Request And What Will We Get For The Funds?

FY 2012 Travel Statistics Program Budget Request ($000)

Program Activity	FY 2010 Actual	FY 2012 Request	Change FY 2010-2012
Travel Statistics Program	2,947	3,064	117
Total[1,2]	$2,947	$3,064	$117

[1] FY 2012 increase reflects reallocation of operating expenses.
[2] The total funding includes salaries and administrative expenses and contract program.

Base Program

The Omnibus Household Survey has been conducted since 2000. Data are collected from over 1,000 nationally representative households and a target sample of 500 households in one of nine selected Metropolitan Statistical Areas. Survey questions cover a variety of areas to better understand the nation's satisfaction with the transportation system including items related to: mode use, journey to work, distracted driving, telecommuting, livable communities, security screening procedures at airports; and bicyclists and pedestrian attitudes and interests. As a result, transportation analysts are better able to identify areas for improvement, develop model parameters for forecasting and establish performance measures to evaluate them. For FY 2012 this survey is scheduled to be conducted on a reimbursable basis with other DOT modes.

The Intermodal Passenger Connectivity Database (IPCD) includes data on facility location and, upon completion, will include all scheduled passenger transportation modes that serve each of the estimated 7,000 passenger terminals in the U.S. This provides an integrated basis for measuring intermodal connectivity in the passenger transportation system, and the only consolidated database of passenger transportation facilities for all modes. The need for intermodal facility data is identified in the USDOT Strategic Plan as necessary for evaluating the state of good repair for our nation's highway system and bridges.

BTS conducts the National Census of Ferry Operators (NCFO) and updates the national ferry database. The NCFO was previously conducted in 2006, 2008 and 2010, and will begin another data collection in 2012 and provide updated information on the nation's ferry operations including information on ferry operators, their vessels, services, and routes. This information will be used to update the BTS database of all ferry operations that provide domestic service within the U.S. and its territories. This includes those operations that provide itinerant, fixed route, common carrier passenger, and/or vehicle ferry service. Information from this database are used in analysis related to emergency routes in the event of natural disasters, safety analysis related to aging vessels, and infrastructure needs in the transportation system. The Federal Transit Administration (FTA) and the Federal Highway Administration (FHWA) are also interested in the data for potential use in formula grant funding.

BTS will continue to provide data users with comprehensive monthly and annual vehicle, container, passenger and pedestrian entry count information for all U.S. land and ferry ports of entry from Canada and Mexico. The data represents activity at the port level on the U.S.-Canadian and U.S.-Mexican land border and international ferry crossings. The data are used for monitoring North American Free Trade Agreement activity, for traffic analysis, and for data-driven resource allocations. These data are also used by the FHWA in calculating apportionments for border state infrastructure grants.

Key Actions:
- Collect data from households (Omnibus Survey) covering issues including air, auto, intercity bus and intercity rail (on a reimbursable basis).
- Maintain and increase the data coverage in the Intermodal Passenger Connectivity Database (IPCD).
- Begin another data collection and provide updated data on the nation's ferry operations; including information on ferry operators, their vessels, services, and routes.

Key Outputs:
- National Census of Ferry Operators results update the BTS database of all ferry operations that provide domestic service within the U.S. and its territories.
- Monthly and annual pedestrians, passengers and vehicle crossings including rail and intermodal crossings into the U.S. from Canada and Mexico (Border Crossing /Entry data) released via the RITA web site in a searchable database. Border Crossing/Entry data will also be made available via www.data.gov.
- Trends in Focus reports using key transportation data with context to timely and relevant travel-related issues.
- Technical reports on travel data and information including data quality issues and data gaps.

Key Outcomes: Update intermodal data on passenger transportation (the Intermodal Connectivity Database) in FY 2012, including refinements to the newly added geographic information system (GIS) interface. This will provide decision makers with the information necessary for planning and investments.

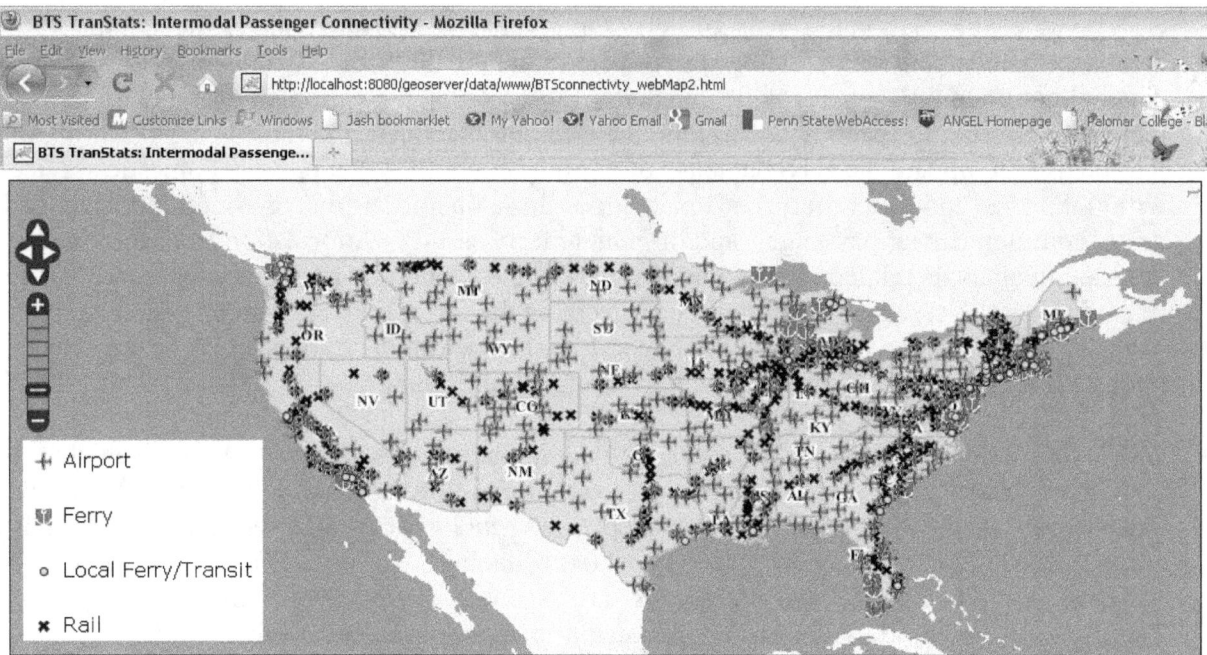

- Release of North American border crossings data by all transport modes for analysis, performance measurement, and congestion research to be used by transportation planners. This fills critical data gaps on the flow of people and vehicles between the U.S. and Mexico and the U.S. and Canada.
- Federal, state and local governments, researchers, educational institutions, transportation decision makers and the general public have intermodal data to understand how people and vehicles move around and through the U.S.

What Is This Program?

Strategic Goals: Safety, environmental sustainability, state of good repair, livable communities, and economic competitiveness.

Purpose/Beneficiaries: Provide data on passenger travel facilities to public and private decision makers.

Partners: USDOT modal administrations, other government agencies, transportation research community and the traveling public.

Description: The Passenger Travel Statistics Program collects, compiles and analyzes passenger travel data for all modes of transportation, and provides multi-modal, analytic reports and products.

Prior Year Accomplishments:
- Updated 2010 NCFO data will support research and analysis, as well as updates to the national database of ferry operators.

- Data on commuter rail stations will be completed and added to the Intermodal Passenger Connectivity Database (IPCD); a special report analyzing the data and comparing commuter rail connectivity with the other modes already analyzed will be issued; and data collection begun during FY 2010 on transit rail stations (heavy rail/subway and light rail) will be completed and added to the database.
- An update to the Rural Access Report, which will include new data on inter-city bus stations, will be published. This will provide information to planners and decision-makers about transportation interconnectivity in rural areas.
- The Border Crossing/Entry Data Program will organize and summarize the U.S. Customs and Border Protection international border crossing and entry data, including comprehensive monthly and annual vehicle, container, passenger and pedestrian entry count information for all U.S. land and ferry ports of entry from Canada and Mexico. This will provide information to decision-makers and planners to target transportation investments.
- Analysis of travel trends with monthly vehicle-miles traveled data; short-run monthly VMT forecasts; analysis and estimation of local VMT; and modeling support and data development of FHWA's Multimodal Transportation Analysis Framework.

Why Is This Particular Program Necessary?

Travel data are critical components of the Intermodal Transportation Data, National Transportation Atlas Database, and Transportation Statistics Annual Report which were previously requested by Congress. Travel data are central to analyses of transportation projects required under the American Recovery and Reinvestment Act, the National Environmental Policy Act, and other federal legislation.

Travel data benefit the American public by assisting with attaining a safe, efficient, economical and sustainable transportation system. Information obtained on passenger travel is needed to identify characteristics of current use of the nation's transportation system, forecast future demand, analyze alternatives for multimodal investment in and development of the system, and assess the effects of federal legislation and federal/state regulations on the transportation system and its use. In addition, economic activity generated from people traveling represents one of the largest industrial sectors in the U.S.

Data relating to this sector are essential to government, private industry, and others responsible for policy formation and promotional activities in the general fields of travel and tourism. The lack of long distance travel data prohibits optimal decision making for transportation investments.

How Do You Know The Program Works?

Effective: BTS data are used to make transportation investment decisions and evaluate system performance. OST, modal administrations, research community, industry analysts and academics, and the general public exhibit high demand for BTS travel data and travel-related research.

Research: BTS routinely receives feedback from its stakeholders and customers through formal and informal means. This includes:
- Customer outreach and listening sessions;
- Input from the online survey (American Customer Satisfaction Index);
- Data requests and downloads from the RITA web pages;
- Participation and membership in customer/stakeholder organizations and associations.

Efficient: Transportation Research Board (TRB) panels have identified these travel data collection programs as core programs to understand passenger travel and demand.

Why Do We Want/Need To Fund The Program At The Requested Level?

History/Outyear Needs: FY 2010 $2.947M; FY 2012 $3.064M.

Detailed Justification for the Freight Statistics Program

What Do I Need To Know Before Reading This Justification?

Understanding how freight moves throughout the country and to and from the nation's ports is essential in informing the USDOT's transportation policy and investment decisions. The Freight Statistics Program is designed to bring this information to light by conducting comprehensive multimodal freight surveys and data analyses.

- Except for temporary setbacks during the recent economic downturn, freight transportation continues a long-term growth, supporting economic activity throughout the US and providing the means for American exports to reach foreign markets.
- Freight transportation is an increasingly significant contributor to congestion, safety exposure, infrastructure performance, and greenhouse gas and other emissions, and local congestion is creating disruptions and added costs for a growing amount of interstate and international commerce.
- The USDOT Strategic Plan identifies the need for better information on freight flows to enhance the nation's ability to make optimal transportation investment decisions.
- The Commodity Flow Survey (CFS) is the only source for national truck shipments and national truck hazardous materials shipment data.
- The International Freight Data System (IFDS) partnership of USDOT agencies supports a requirement of Section 405 of the SAFE Port Act of 2006.
- BTS works with USDOT agencies with regulatory authority over hazardous materials transport in the CFS effort to ensure the CFS effectively addresses emerging issues in this area.

What Is The Request And What Will We Get For The Funds?

FY 2012 Freight Statistics Program Budget Request ($000)

Program Activity	FY 2010 Actual	FY 2012 Request	Change FY 2010-2012
Freight Statistics Program	10,723	16,021	5,298
TOTAL[1]	**$10,723**	**$16,021**	**$5,298**

[1] The total funding includes salaries and administrative expenses and contract program.

Base Program
The base program includes continued analysis and research using the 2007 Commodity Flow Survey data and international trade and freight transportation data and analysis, including data for the formula used in calculating apportionments for border state infrastructure grants (Federal Highway Administration). The Freight Statistics Program produces special and focused reports on key transportation issues.

Commodity Flow Survey

The Commodity Flow Survey (CFS) is the flagship survey of the Freight Statistics Program and has been recognized within USDOT and by external customers as one of the USDOT's most valued freight data series since its initiation in 1993. Although the CFS is a multimodal survey, it is the only available source of national freight data for the highway mode. It also provides the most comprehensive set of national data on the movement of hazardous materials. These data are also fundamental in supporting the USDOT's strategic objective of economic competitiveness.

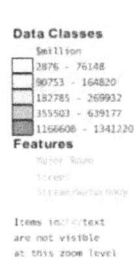
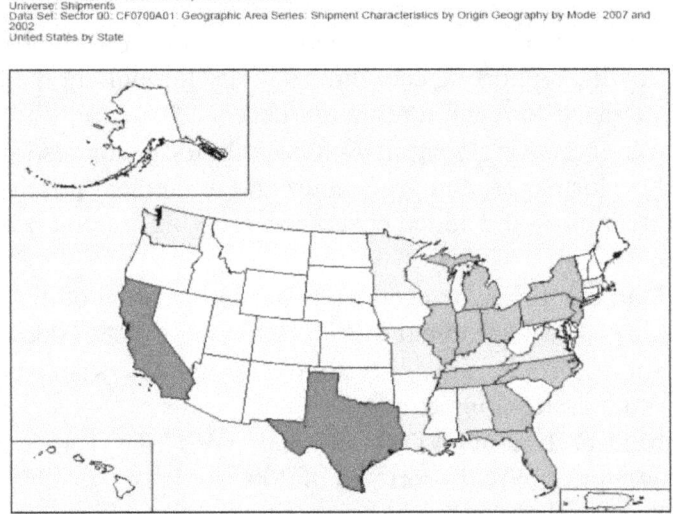

International Data Program

BTS is currently focusing its International Data Program on data collection, analysis and outreach. Topics covered by the current program include U.S. international trade and transportation, travel and border crossings, North American transportation research, and global competitiveness. All of these functional areas are focused on data collection, integration, analysis and outreach. The largest topic area for the International Data Program is international multi-modal freight transportation data and information. Transborder data collection and analysis is ongoing and is released monthly. The International Freight Data System (IFDS) partnership project is DOT agencies' interface with the Customs and Border Protection's International Trade Data System (ITDS). Section 405 of the SAFE Port Act of 2006 formally established ITDS. The SAFE Port Act requires all agencies with import and export regulations and authorities to participate in ITDS.

Key Actions:

- Release monthly the North American freight transportation (Transborder Freight) data for customers including USDOT staff, Congressional staff, state DOTs, international organizations, universities and academics, and others.
- At least one value-added report will be developed in early 2011 – "U.S. Freight Movement Highlights" utilizing 2007 CFS and other freight data sources.

- Conduct research and analysis using available freight transportation data to support USDOT's strategic goals and to inform on key transportation issues and events that impact all modes of transportation.

Key Outputs:
- Comprehensive and timely multimodal and intermodal import and export data are available.
- Publication of the international and intermodal freight reports provide data profiles of the nation's leading international and intermodal transportation gateways. It highlights key gateways and border crossings for air, water, and surface modes.
- Hazardous materials import data delivered to USDOT modal administrations. These data fill critical data gaps, improve risk assessments and support evaluations of entities that transport hazardous materials improving the safety of the U.S. transportation system.

Key Outcomes:
- Comprehensive 2007 CFS-related data products and other freight data products and research easily accessible online for transportation decision makers and the general public.
- Customers use Transborder Freight data for a variety of purposes, including trade corridor studies, risk assessments, decision making, transportation infrastructure planning, and other purposes.
- Improved understanding the demand that freight movements place on the U.S. transportation system.

Program Funding Increases:

Commodity Flow Survey (CFS) $2M

Even though it is considered the most comprehensive national survey of commodity movements, limitations to CFS data are caused by sample size constraints and coverage issues. In addition, the CFS scope and methodology must be continuously assessed and reexamined to address changes in freight transportation logistics and business operations. The 2012 CFS will oversample specific industries/geographies to provide more robust estimates for specific modes (particularly less frequently used modes like water and air) or specific important commodities (e.g., biofuels).

The FY 2012 enhancement request is for $2M to maintain the current sample size and operation and support a slightly expanded coverage of establishments.

The Vehicle Inventory and Use Survey (VIUS) $3M

The $3M funding increase for VIUS will capture the physical and operating characteristics of trucks nationally—crucial data for evaluating the role of trucks in transportation. Data from the VIUS will be used to conduct safety analysis, estimate fuel consumption and the economic productivity of trucks, and develop statistics of highway usage and cost allocation. Since a significant percentage of trucks are owned by households and used for personal transportation, the VIUS will collect key information about passenger travel as well as freight. Plans include expansion of coverage to automobiles and buses for a very small cost increment to provide a complete picture of vehicle use for personal and business purposes.

International Freight Data System (IFDS) $.9M

This increase to the IFDS program will be used to address safety, regulatory, legislative and policy issues, and to conduct research and analysis. International data used by USDOT are obtained from this and other sources and disassociates the shipper from the manifest (or transportation) data, which impedes the production of accurate transportation statistics. USDOT obtains international transportation data from Customs and Border Protection (CBP).

The IFDS would improve import and export data quality and provide an automated system for obtaining multimodal and intermodal international cargo movement from a single source. USDOT modal partners will obtain critical data including hazardous materials shipment data, vehicle and vehicle parts import data, mode of transportation data, and analytical and statistical data via the IFDS to support resource allocation decisions, risk assessments, licensing decisions, and safety standard reviews. This project also supports agencies with import and export (global supply chain security) roles and responsibilities.

RITA aims to design and build the IFDS data warehouse and interface with CBP on behalf of USDOT modal administrations, which will share multi-modal data from a single portal.

Ongoing and consistent commitment for annual contributions from the modal administrations for operation and maintenance (O&M) is required due to budget constraints. To alleviate a potential risk for recurring O&M costs, this request would fund O&M for the IFDS on behalf of the USDOT partnering agencies.

Key Actions: Conduct an enhanced/expanded 2012 Commodity Flow Survey (CFS) data collection; operate and maintain the International Freight Data System; design the Vehicle Inventory and Use Survey (VIUS).

Key Outputs: Complete 2012 CFS data collection; Methodology for the VIUS and targeted surveys; IFDS data delivery to modal customers.

Key Outcomes: Comprehensive national and international freight data made available for transportation planning and policy decisions for all modes of transportation. Data will be available to transportation decision makers to support data-driven decisions, to support efficient allocation of resources and effective transportation investments.

What Is This Program?

Strategic Goals: Safety, environmental sustainability, state of good repair, livable communities, and economic competitiveness. The USDOT Strategic Plan identifies the need for intermodal and cross-modal research and data-driven decision-making. For example, through these programs, national (CFS) and international (IFDS) hazardous materials shipment data will be made available for the USDOT agencies for safety, regulatory evaluation, analysis, and risk assessment.

Purpose/Beneficiaries: Develops and compiles data on freight movements within, through, into and from the U.S. by all modes of transportation.

Partners: USDOT modal administrations, other federal agencies and the transportation industry.

Description: Collects, compiles, and analyzes freight data for all modes of transportation, and provides analytic reports and products from a multimodal and intermodal perspective on the performance and impacts of national and international freight flows on the national transportation system.

Prior Year Accomplishments:
- Complete research efforts and be fully engaged in the planning and implementation of the 2012 CFS.
- BTS will sponsor a CFS Data Users Conference.
- Complete 2007 value-added data products.
- Execute a release schedule for the TransBorder Freight Data.
- Collaborate with the U.S. Census Bureau, Canadian and Mexican transportation and statistics agencies, and other federal agencies in the North American Transportation Statistics Interchange.
- Complete trilateral update of the North American Transportation Statistics (NATS); lead the design and release a summary report on key trilateral indicators based on NATS-OD (online database).
- Continue to conduct outreach to customers to assure that the most important data are provided, and develop innovative methods for meeting freight data needs at the state and local levels.
- Lead the first phase deployment of the IFDS data warehouse for USDOT, in partnership with the Department of Homeland Security (DHS); the IFDS is the USDOT interface to the DHS Customs and Border Protection's International Trade Data System (ITDS).
- Prepare reports on key transportation issues such as container ports, NAFTA trade and freight transport, trade with U.S. trade partners, national freight movements (including hazardous materials shipments), and changes in global transportation activities.

Why Is This Particular Program Necessary?

Information on the movement of goods as described in the Intermodal Transportation Data, National Transportation Atlas Database, and Transportation Statistics Annual Report has been requested by Congress in the past. Goods movement data are central to analyses of transportation projects required under the American Recovery and Reinvestment Act, the National Environmental Policy Act, and other federal legislation.

Support of interstate commerce is a core function of the federal government. Data describing the volume, value, and geography of freight movement and performance of the transportation system identify the most cost-effective investments in freight infrastructure. These data help policy-makers understand the potential effectiveness of proposed regulations, financial incentives, and other policies to improve safety and environmental sustainability. Freight data are also key to understanding and enhancing the role of the domestic transportation system in moving U.S. international trade, particularly in supporting the Presidential goal of increasing exports.

Further expansion of the 2012 CFS will allow for increased flow data and better commodity by mode estimates. Through the first phase of the IFDS data warehouse and interface with Customs and Border Protection's import data, USDOT will obtain international transportation data to address safety, environmental and policy issues, and for research and analysis.

Economic activity is variable and must be monitored to assure that the freight transportation system is responding effectively to the nation's logistical needs. Failure to meet the logistical needs of manufacturing and services can result in loss of jobs and reduced spending on other economic activities. Without requested FY 2012 funds, the CFS would not continue and the national freight data program would decline.

How Do We Know The Program Works?

Effective: Demand from the USDOT, modal administrations, research community, industry, state and local governments, and the general public. A wide array of transportation data users rely on data from the Freight Data Program, including researchers and media who utilize the data and analysis.

Research: BTS receives routine feedback from its stakeholders and customers through formal and informal means. This includes:
- Customer outreach and listening sessions;
- Input from the American Customer Satisfaction Index;
- Data requests and downloads from the RITA web pages;
- Participation in freight-related conferences and workshops to proactively seek feedback on our products.

Sub-program balance: Sub-programs necessary for understanding freight flows.

Efficient: TRB panels have identified these collections as the optimal approach.

Why Do We Want/Need To Fund The Program At The Requested Level?

Funding options: FY 2012 $16.021M

History/outyear needs: FY 2010 $10.723M; FY 2012 $16.021M.

Detailed Justification for the Transportation Economics Program

What Do I Need To Know Before Reading This Justification?

The U.S. transportation system forms the backbone of the nation's economy. To better understand this relationship and maximize economic competitiveness, the Transportation Economics Program develops and analyzes data that helps to explain the economic impact of the USDOT's transportation policy and investment decisions.

- The Transportation Economics program supports the USDOT economic competitiveness strategic goal and provides key information for understanding the consequences of user fees and taxes.
- Congress has previously mandated that BTS compile and publish statistics supporting economic and transportation analysis for transportation decision-making by all levels of government, the transportation community, businesses and consumers.

What Is The Request And What Will We Get For The Funds?

FY 2012 Transportation Economics Budget Request ($000)

Program Activity	FY 2010 Actual	FY 2012 Request	Change FY 2010-2012
Transportation Economics Program	1,811	1,938	127
TOTAL [1,2]	$1,811	$1,938	$127

[1] FY 2012 increase reflects reallocation of operating expenses.
[2] The total funding includes salaries and administrative expenses and contract program.

Key Actions:
- This program compiles Government Financial Transportation Statistics (GTFS) and the Transportation Satellite Account (TSA), and generates economic impact estimates using multi-modal transportation models.
- The Transportation Satellite Account expands upon the basic Bureau of Economic Analysis (BEA) accounts which describe the for-hire transportation industry by providing coverage of in-house transportation as well. Detailed inputs of services and commodities by mode are provided by The Transportation Satellite Account for in-house transportation, supplementing the data that BEA accounts provide on transportation.
- Develops basic economic and financial data to support transportation decision-making, including investment analysis, through development of economic indicators that measure the contribution of transportation to the economy, research on transportation economic trends and forecasts, transportation costs and expenditures.

Key Outputs:
GTFS; updated Transportation Satellite Account; and economic analysis and modeling; data and findings used to assess transportation impacts on the U.S. economy.

Key Outcomes:
Provide policy-makers with greater insight into the impact of transportation on the economy and Gross Domestic Product (GDP).

What Is This Program?

Strategic goals: Economic competitiveness and environmental sustainability.

Purpose: Transportation and related investments are critical to the competitiveness of the U.S. economy and represent significant investments by governments and households. This request funds the continuation of the Transportation Economics Program and enhancements to the BTS Economic and Competitiveness Data and Analysis Program for producing the annual GTFS report to address government transportation finance and expenditures; update the TSA, which measures the role of transportation in the economy; and develop and use economic models for policy and economic impact analysis.

Beneficiaries: OST, FHWA, FAA, Congress, state and local governments.

Partners: BTS works with FHWA to support their modeling and forecasting needs; the Bureau of Economic Analysis and state governments provide critical data.

Description: Program products provide transportation officials with information on the economic impact of transportation and how to optimize transportation investments, improve system productivity, and increase the value of transportation to users. Specifically:
- The GTFS report provides the only single source of statistics on transportation-related revenues and expenditures of the federal, state and local governments for all modes of transportation. The report also contains federal budget authority and obligations, and grants to state and local governments. Statistics on federal expenditures, budget authority and obligations are provided at the agency and program level.
- The TSA expands upon the basic Bureau of Economic Analysis (BEA) input-output (I-O) accounts, which describe all industries and the commodity inputs used in production, by providing separate coverage of for-hire and in-house transportation and detailed inputs of services and commodities used in the transportation sector by mode. These comprehensive estimates for the transportation sector are not available from BEA.
- The BTS Trending and Forecast Team performs trending, prediction, seasonal adjustment, and time series and econometric forecasting of transportation data. The team works with federal transportation forecasters outside of BTS, as well as with the modal agencies across USDOT, on indexes, VMT forecasting, and modeling. This assists policy makers in identifying trends and turning points in the economy.

Prior Year Accomplishments:
The Transportation Economics Program will:
- Update the most recently available 2002 TSA estimates for private truck, rail, aviation, and waterborne modes for use by the USDOT Chief Economist, FHWA Office of Policy Studies, FAA, AASHTO and other economists to measure impacts of transportation expenditures on GDP.

- Produce the quarterly Air Travel Price Index (ATPI); continue development of new ATPI automated system components, and user testing; develop consumer choice demand component to ATPI in order to measure consumer shifting between long and short distances, and high and low cost carriers.
- Continue development of a Time Series and Forecasting Center and publish time series trends, including publication of Transportation Trends in Focus (TTiF) reports on requested topics. The TTiF series reports on key transportation data.
- Provide aviation and other modal data, modeling and forecasting expertise to the Federal Highway Administration in constructing multi-modal network and forecasting model.
- Develop estimates of short-term time series and mid-term TEAMS model travel forecasts; provide support to FHWA on forecasting of VMT and trends in historical nationwide congestion measures; estimate de-seasonalized monthly travel data for FHWA.
- Make estimates of the economic impacts on the Highway Trust Fund of changes in gasoline prices and total number of vehicle miles traveled. Add travel and program modeling support to modal agencies through development of the TEAMS model for various modes of transport.
- Publish the 2010 edition of the Government Transportation Financial Statistics (GTFS).

Why is this Particular Program Necessary?

Information on economic performance and an accounting of capital stocks is explicitly required for the Intermodal Transportation Data and Transportation Statistics Annual Report have been requested by Congress in the past.

The program provides unique products to USDOT and other policy makers to estimate how changes in the transportation sector impact the economy. It provides a comprehensive accounting of transportation revenues and expenditures across all levels of government and throughout the private sector portion of the economy. Coverage is essential to understand the consequences of changes to user fees and taxes, as well as business and household logistics costs. This is needed for understanding the potential impacts of changes in energy markets and policies on the transportation sector and other parts of the economy that depend on transportation.

The TSAs will provide data to link project and programs from transportation to the economy for cost-benefit purposes, financial, investment and program evaluation. The TEAMS product will estimate various scenarios for policy development and evaluation, investment analysis, financial planning, program evaluation and budget impacts.

Without this funding, USDOT would not be able to calculate the impact of transportation expenditures on the economy—critical for measuring economic competitiveness.

How do You Know the Program Works?

Effective: Frequency of use of products and information, and inquiries from the White House, Congress, and OST. The program generates unique products that are widely used by federal, state and local government agencies as well as university researchers.

Research: The TSA is based on methodologies developed in partnership with the Bureau of Economic Analysis.

Efficient: This program is useful to planners and policy makers for assessing the effectiveness and efficiency of the transportation system and for evaluating transportation investments.

Why Do We Want/Need to Fund the Program at the Requested Level?

The requested level of funding is needed to continue the base program.

History/Outyear Needs: FY 2010 $1.811M; FY 2012 $1.938M.

Detailed Justification for the Geospatial Information Systems (GIS) Program

What Do I Need to Know Before Reading This Justification?

Much of the data in transportation is best represented on a map. The Geospatial Information Systems (GIS) program develops transportation maps that display this data to support USDOT agencies, the USDOT Crisis Management Center, Congress, and the transportation community. Analysis of the geospatial transportation data is also done by the GIS program to support USDOT policy and investment decisions.

- The GIS Program has been Congressionally mandated in the past, and annually develops, maintains, and distributes the National Transportation Atlas Database (NTAD), a set of nationwide geospatial databases of transportation facilities, transportation networks and associated infrastructure.
- USDOT's Geospatial Information Officer resides within BTS.

What Is The Request And What Will We Get For The Funds?

FY 2012 Geospatial Information Systems Program Budget Request ($000)

Program Activity	FY 2010 Actual	FY 2012 Request	Change FY 2010-2012
Geospatial Information Systems Program	1,758	1,521	-237
TOTAL[1,2]	$1,758	$1,521	-$237

[1] FY 2012 decrease reflects reallocation of operating expenses.
[2] The total funding includes salaries and administrative expenses and contract program.

Key Actions::
- Provide standards and technical expertise to USDOT.
- Distribute transportation data for mapping applications.
- Operate the USDOT ARRA web mapping application.
- Develop the transportation layer of the National Spatial Data Infrastructure (NSDI) as established under Office of Management and Budget (OMB) Circular A-16.
- Represent the USDOT in the transportation geospatial community at conferences and committees and provide a critical nexus for information and data sharing.

Key Outputs:
- Produce and distribute the 2012 edition of the National Transportation Atlas Databases.
- Continue to provide geospatial mapping, analysis, and application development services to Congress, RITA, the USDOT Crisis Management Center (CMC) and other USDOT modal administrations.

- Continue to provide geospatial mapping, analysis, and application development services to Congress, RITA, the USDOT Crisis Management Center (CMC) and other USDOT modal administrations.

Key Outcomes:
- Improved methods for understanding transportation activities patterns enabled through geographic identification of critical characteristics of livable communities, infrastructure weaknesses, and safety risks.

What Is The Program?

Strategic Goals: Safety, livable communities, environmental sustainability, state of good repair, and economic competitiveness.

Purpose: The GIS program provides the USDOT Geographic Information Officer, leads USDOT in the use of geospatial representation of transportation information and data, and partners with the Environmental Protection Agency (EPA) and FHWA to maintain a geospatial dataset of EPA non-attainment areas to support USDOT environmental sustainability strategic goal.

Beneficiaries: The GIS program provides geospatial mapping, analysis, and application development services to RITA, USDOT Crisis Management Center (CMC), other USDOT agencies, and Congress. Cartographic products are used in publications; provided to the CMC to support their exercises and emergency situations; and are submitted to Congress.

Partners: EPA, FHWA, Data.gov, VisualDOT, and the CMC.

Program Description: The GIS program develops transportation geospatial data applications to support USDOT agencies, the CMC, Congress, and the transportation community. Program staff performs geospatial analyses to aid USDOT decision-makers. The program is leading efforts to develop web mapping tools that will monitor a range of sustainability metrics of U.S. cities. The GIS program has also developed geospatial applications to aid in the visualization and analysis of transportation data, including a desktop application to estimate multi-modal trip distances traveled by freight shipments and a web application to display the transportation projects funded by the American Recovery and Reinvestment Act (ARRA).

Prior Year Accomplishments:
The GIS program will:
- Produce and distribute the 2011 NTAD;
- Develop new geospatial datasets for rail bridges and ferry routes;
- Produce 2011 edition of *Major Transportation Facilities of the United States* map;
- Support the CMC during National Level Exercise 2011 and during emergency situations;
- Produce maps for USDOT publications;
- Operate and maintain the USDOT ARRA web mapping application;
- Develop new web mapping applications that showcase USDOT data;
- Represent the USDOT in the FGDC, including chairing the Transportation Subcommittee;
- Assist planning of AASHTO GIS for Transportation Symposium; and

- Continue the partnership with NSGIC and complete the strategic plan for the "Transportation for the Nation" initiative.

Why Is This Particular Program Necessary?

The GIS program is responsible for the development of the transportation layer of the National Spatial Data Infrastructure (NSDI) as established under Office of Management and Budget (OMB) Circular A-16.

The GIS program annually develops, maintains, and distributes the National Transportation Atlas Database (NTAD), which has been requested by Congress in the past.

The GIS program develops the transportation layer of the NSDI and annually distributes the NTAD. This program leads the USDOT effort to develop a web mapping tool to evaluate the sustainability of U.S. cities.

Without this funding, extensive transportation mapping data would not be available. USDOT would not have access to a standardized inventory of geospatial data necessary to map locations of critical transportation infrastructure. USDOT relies on the GIS program to provide geospatial data services related to the ARRA implementation.

How Do You Know The Program Works?

Effective: Customers receive high quality geo-spatial products from the GIS program. More than 2,000 NTAD DVDs are distributed annually. The NTAD download site receives more than 20,000 annual visits. During the past decade, the GIS Program has produced more than 60 special maps per year. These maps enhance USDOT publications and support the CMC.

Research: There is consistent high demand for products and services. The NTAD has been produced every year since 1995. Each year, over 2,000 DVDs are distributed. The American Customer Satisfaction Index web survey on the BTS site also captures stakeholder feedback on GIS products and services. BTS stakeholder outreach has revealed a desire for centralized GIS reporting at the USDOT and the GIS program provides this service for USDOT.

Efficient: Recent stakeholder outreach found the GIS program offers substantial value for a low-cost investment as evidenced by customer demand.

Why Do We Want/Need To Fund The Program At The Requested Level?

Funding Options: $1.521M would fund the base program.

History/Outyear Needs: FY 2010 $1.758M; FY 2012 $1.521M.

Detailed Justification for Transportation Analysis, Data Quality, and Performance Metrics

What Do I Need To Know Before Reading This Justification?

Access to high quality data and transportation metrics is vital in understanding the state of the transportation system. The Transportation Analysis, Data Quality, and Performance Metrics program works to ensure that the quality of U.S. DOT data is of the highest standards and that the metrics used to measure transportation performance are comprehensive.

- BTS provides a wide range of transportation data analysis, such as the Transportation Services Index (TSI) for decision-makers and the public.
- Within BTS the Office of Statistical Quality provides technical support to the Administration's performance management agenda, which includes the development of more comprehensive and outcome-focused performance measures.

What Is The Request And What Will We Get For The Funds?

FY 2012 Transportation Analysis, Data Quality, and Performance Metrics Budget Request ($000)

Program Activity	FY 2010 Actual	FY 2012 Request	Change FY 2010-2012
Transportation Analysis, Data Quality, and System Performance	7,416	7,268	-148
TOTAL [1,2]	$7,416	$7,268	-$148

[1] FY 2012 decrease reflects reallocation of operating expenses.
[2] The total funding includes salaries and administrative expenses and contract program.

Key Actions:
- Produce, publish and make available online and limited print documents, and scheduled data releases. Provide technical support and transportation expertise to international data partnerships/exchanges.
- The program will continue to produce relevant and timely focused analytical and technical reports on multimodal and intermodal transportation and statistics-related issues.
- Provide data and statistics to agencies and organizations within and outside the U.S. in support of national and international statistics compilations.
- Serve as the lead USDOT participant in the North American Transportation Statistics Interchange to promote and develop high quality, relevant, comparable data, and analysis for an efficient and fully integrated transportation system for North America.
- Ensure the quality of the statistical compilations and of individual data programs throughout USDOT, and develop and maintain standards for transportation data to ensure accuracy and reliability.

- Support USDOT in the interpretation and implementation of the statistical portion of the USDOT Information Dissemination Quality Guidelines.

Key Outputs:
- 2011 Pocket Guide to Transportation;
- 2011 Transportation Statistics Annual Report;
- Monthly scheduled releases of the Transportation Services Index (TSI);
- Bi-monthly updates and releases of Key Transportation Indicators, including development of new indicators;
- Quarterly scheduled updates of the National Transportation Statistics (NTS);
- Guidance to OST and modal administration on performance measures; BTS reports; and quality standards.

Key Outcomes:
- Improved data quality and reliability;
- Improved online access to key transportation statistics and analyses;
- Online availability of intermodal and multimodal transportation data; and
- Reliable performance measures for USDOT programs and projects.

What Is This Program?

Strategic Goals: Safety, economic competiveness, livable communities, state of good repair, environmental sustainability, and organizational excellence.

Purpose/Beneficiaries: This program makes high quality transportation data for all modes easily accessible. This program also develops transportation data standards and improves data accuracy.

Partners: USDOT modal administrations, OST, OMB, and international organizations.

Description: The Transportation Analysis, Data Quality and Performance Measurement program performs research and publishes multi-modal and intermodal transportation data and analysis covering critical and timely transportation topics with the goal of providing quality data and information for all modes of transportation for decision-making.

Prior Year Accomplishments:
- Support the USDOT in the interpretation and implementation of the statistical portion of the USDOT Information Dissemination Quality Guidelines.
- Support requests from USDOT, such as review of statistical methods in rulemaking processes to ensure that rulemakings are based on sound data.
- Conduct statistical data quality reviews of data compilations and products.
- Provide statistical support for the USDOT's Performance and Accountability Report.
- Produce and publish the *Transportation Statistics Annual Report* (TSAR).
- Produce and release the monthly Transportation Services Index (TSI).
- Update, publish, and distribute the annual *Pocket Guide to Transportation*.

- Update the web compendium of the *National Transportation Statistics* quarterly and produce an associated volume of the *State Transportation Statistics*.
- Bi-monthly web update of the Key Transportation Indicators; evaluate Key Transportation Indicators Project for expansion.
- Produce relevant and timely focused analytical and technical reports on transportation and statistics-related issues and data.

Why Is This Particular Program Necessary?

Information on performance of the transportation system for the Transportation Statistics Annual Report and guidelines for improved data quality and review of statistical reliability have been requested by Congress in the past.

This program supports the emphasis on performance measurement and management to improve the delivery of transportation services. Key to effective performance measurement is the availability of reliable, timely data, and the presentation of performance measures in effective forms.

This program brings together data from multiple sources to produce the *Transportation Statistics Annual Report*, which presents the state of the transportation system and the state of transportation data.

Statistical capabilities are necessary to ensure the reliability and validity of the statistical agency's products, support the data needs of USDOT, and facilitate compliance with legal requirements of the Information Quality Act. BTS evaluates performance measures used to assess USDOT performance.

Without this funding, the White House, Congress, OST and USDOT modal administrations would not have the multi-modal and intermodal data and information they now use for comparative analysis. USDOT would have no centralized authority and expertise in data quality standards and performance measurement.

How Do You Know The Program Works?

Effective: High demand from OST, modal administrations, researchers, and the public.

Research: Customer outreach; online customer survey (American Customer Satisfaction Index); data requests and downloads from RITA web pages and the library reference services; and stakeholder feedback.

Sub-program Balance: Emphasis on data quality, performance measurement; data dissemination and international partnerships to cover critical areas.

Efficient: One organization for transportation expertise and data quality guidance in all modes.

Why Do We Want/Need To Fund The Program At The Requested Level?

Funding options: FY 2012 $7.268M

History/outyear needs: FY 2010 $7.416M; FY 2012 $7.268M.

Detailed Justification for the National Transportation Library

What Do I Need To Know Before Reading This Justification?

The National Transportation Library maintains and facilitates access to statistical and other information needed for transportation decision-making at the federal, state, and local levels. Transportation information requests are also facilitated through the National Transportation Library for Congress, industry, the media, and the public.

What Is The Request And What Will We Get For The Funds?

FY 2012 National Transportation Library Budget Request ($000)

Program Activity	FY 2010 Actual	FY 2012 Request	Change FY 2010-2012
National Transportation Library	2,345	2,139	-206
TOTAL [1,2]	$2,345	$2,139	-$206

[1] FY 2012 decrease reflects a reallocation of operating expenses.
[2] The total funding includes salaries and administrative expenses and contract program.

Base Program
Key Actions:
- Enter into new partnerships with transportation information providers.
- Improve information access tools and user interfaces.
- Continue digitization of historical transportation information.
- Continue maintenance of transportation information standards.
- Coordinate the activities of the national transportation knowledge network.

Key Outputs:
- Updates to transportation information management standard, such as the Transportation Research Thesaurus (TRT).
- New mobile applications providing access to transportation information.
- New formats of information included in library collections, such as images and video.
- New information resources, databases, and collections included in NTL information access portals/searches.

Key Outcomes:
- Increased access to information for USDOT and other transportation decision makers.
- Increase in users of NTL resources over FY 2011 levels.
- Decrease in redundant research throughout the transportation community.
- Increased productivity of Departmental staff and transportation professionals.

What Is This Program?

Strategic Goals: Safety, livable communities, environmental sustainability, state of good repair, and economic competitiveness.

Purpose/Beneficiaries: NTL maintains and facilitates access to statistical and other information needed for transportation decision-making at the federal, state, and local levels. NTL also supports the professional development competencies and skills to transportation information professionals. / Congress, the media, researchers, transportation professionals, federal, state, and local governments, USDOT staff, and the general public

Partners: Transportation Research Board (TRB), American Association of State Highway and Transportation Officials (AASHTO), public and private transportation libraries and information providers, library networks and organizations.

Program Description: The NTL operates in four functional areas:
Reference Management NTL reference services are the transportation information front door to the USDOT. The NTL Reference Service team handles 1,500-2,500 requests for information each month, provides a searchable FAQ database used between 60,000-80,000 times per month, and delivers training and outreach.

Database and Archive Management: The NTL Integrated Search platform includes a Digital Repository, web portal, and single search interface for transportation information resources. Through the search interface, users can search all NTL and other transportation resources at once. In FY 2009, an average of 108,557 individual users accessed NTL web resources and tools 202,962 times every month. With the FY 2012 NTKN enhancement, the number of databases managed will increase to include maintenance of a portal for transportation datasets and development and maintenance of an expertise system.

Tools and Standards Management: The NTL's metadata standard for indexing digital resources and use of controlled vocabularies (e.g., Transportation Research Thesaurus, or "TRT") allows interoperability with other web resources and targeted access to transportation information resources. NTL maintains the TRT, the international standard transportation taxonomy and controlled vocabulary. NTL also increases access to information through implementation of Google Sitemaps and other protocols, joining science.gov, a joint project of 17 federal agencies providing access to federal scientific and technical information, and maintaining access tools such as directories, bibliographies, and custom Google searches of all state DOT, UTC, MPO, and transit agency websites. These tools and standards are used in tandem with the databases and archives enabling efficient, robust search, retrieval, and access to transportation information.

Networking: NTL's networking activities include collaborative efforts to develop and provide improved access to information and support the professional development of the information management community. Existing tools and activities include the Transportation Libraries Catalog, Transportation Librarians Roundtable, and coordination of the three regional Transportation Knowledge Networks (TKNs) toward development of NTKN.

Prior Year Accomplishments:
- NTL support of USDOT TIGER Team efforts on the ARRA.
- Provision of quick, courteous and accurate answers to requests for information from Congress, federal agencies, academia, industry, the media, and the public within 24 – 48 hours of receipt.
- Develop, maintain, and promote new and existing tools and standards.
- Coordination of national information dissemination activities through the NTKN.

Why Is This Particular Program Necessary?

Congress previously authorized BTS to establish the NTL to collect, maintain, and facilitate access to transportation data and information and authorized NTL to work with partners in the transportation information community in the pursuit of this goal.[1] These activities support the Open Government Initiative goal of transparency to increase access to government information. The proposed Federal Research Public Access Act (FRPPA) would require any published research funded in full or in-part by federal dollars be deposited into a repository at the sponsoring federal agency. The Consolidated Appropriations Act of 2008 included this provision for the National Library of Medicine (NLM). FRPPA would expand the requirement to all Federal research dollars. NTL's Digital Repository infrastructure and processes are similar to that of NLM's PubMed database and can be adapted to fulfill this potential requirement for the Department.

The NTL serves as a one-stop portal providing access to core information resources; access to local, regional, national, and international resources; improved services, protocols, and standards to facilitate information sharing and improve professional competencies. The NTL program supports the information needs of USDOT stakeholders and the general public, as well as the transportation information professionals.

Without the NTL program, no economies of scale achieved by consolidation of commercial databases and use of a single technological infrastructure for USDOT-generated databases will be realized. Access to library programs and services saves time and money, and directly supports decision-making.[2]

How Do You Know The Program Works?

Effectiveness: High and increasing demand from OST, modal administrations, researchers, and the public evidence the NTL program's effectiveness. For example, an increase from 58 (FY 2008) to 63 (FY 2009) average daily visitors to the physical library or a 50% increase in usage of an electronic database from December 2008 to April 2010, demonstrate that resources and services NTL provides are increasingly useful for a larger proportion of users.

---

[1] Safe Flexible Efficient Transportation Equity Act: A Legacy for Users, 49 U.S.C. § 111 (f) (2006)
[2] Roger Strouse, *Information Management Under Fire: Capturing ROI for Enterprise Libraries*, Information Management Service Briefing, volume 10, November 2007.

Research: In addition to the SR284 and NCHRP 643 reports, BTS customer outreach, data requests, and NTL independent analyst activities inform the program's operation. BTS uses the American Customer Satisfaction Index web survey to evaluate customer satisfaction with overall and specific features of the website. In May 2010, BTS engaged in listening sessions with 23 major stakeholders, which includes feedback on the website and program. All feedback is tracked and analyzed according to GAO recommendations. In FY2010, the NTL program's technology infrastructure and plan will be reviewed by independent information analyst firm Outsell, Inc./Gilbane Group.

Efficiency Measures: Community-wide adoption of standardized information storage and retrieval procedures reflects NTL's efficiencies. The program achieves this goal through strategic resource use and encouraging transportation information partners' use of standards and protocols.

Why Do We Want/Need To Fund The Program At The Requested Level?

The requested level of funding is needed to continue the base program.

Funding Options: FY2012 $2.139M

History/outyear needs: FY 2010 $2.345M; FY 2012 $2.139M.

THIS PAGE HAS BEEN INTENTIONALLY LEFT BLANK

Detailed Justification for the Airline Transportation Statistics Program

What Do I Need To Know Before Reading This Justification?

The Airline Transportation Statistics Program provides comprehensive data and analysis to support U.S. DOT policies, programs, and regulations regarding the airline industry. Air transportation plays a critical role in enhancing the economic competitiveness of the nation and it is the data provided by the Airline Transportation Statistics Program that enables the U.S. DOT to make decisions that are well informed.

- New USDOT rulemakings have expanded the types of on-time performance data that large airlines must report, including tarmac delay times.
- Assuming the June 8, 2010, Notice of Proposed Rulemaking entitled Enhancing Airline Passenger Protections is adopted and final in 2011 (becoming effective in 2012); additional resources are needed for processing additional on-time data that would be submitted by additional carriers.
- Recent accomplishments include the initiation of a new collection of tarmac time data and its dissemination through the Air Travel Consumer Report and BTS website.
- Funding for this program is provided by the Federal Aviation Administration (FAA) through an interagency agreement.

Number of Large Carrier Flights with Tarmac Delays of More than Three Hours, by Month

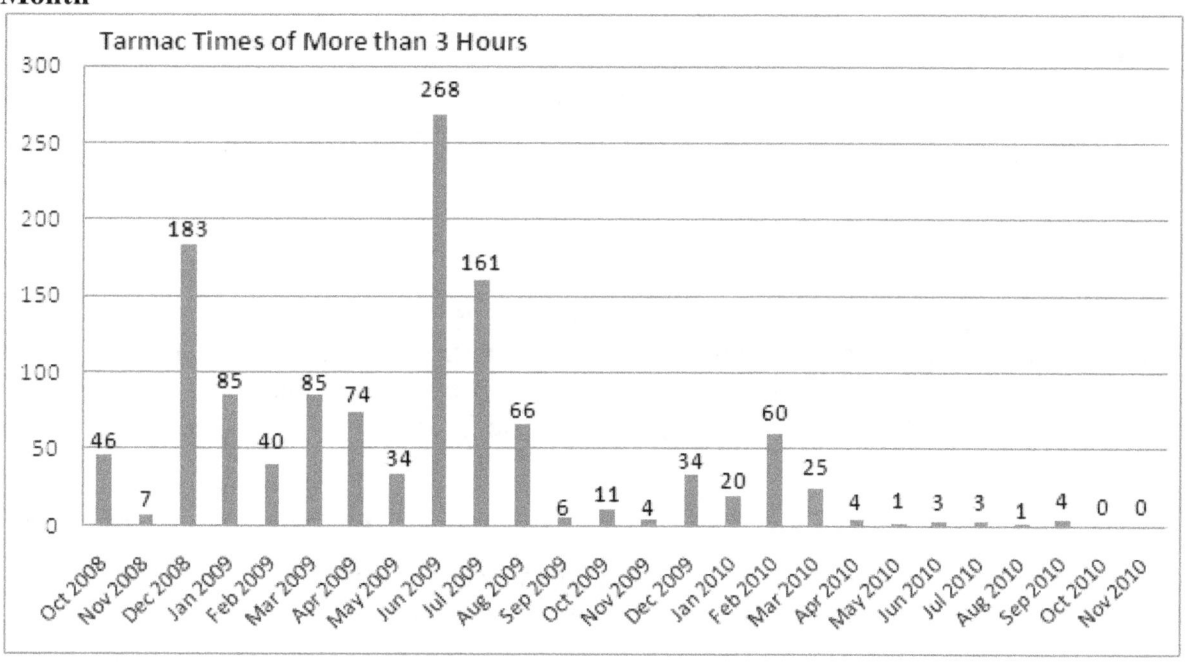

What is the Request and What Will We Get for the Funds?

FY 2012 Airline Transportation Statistics Program
Budget Request
($000)

Program Activity	FY 2010 Actual	FY 2012 Request	Change FY 2010-2012
Airline Transportation Statistics Program	[4,000]	[5,000]	[1,000]
Total[1]	[$4,000]	[$5,000]	[$1,000]

[1] The total funding includes salaries and administrative expenses and contract program.

Key Actions:
- Maintain national airline data system: collect and disseminate airline financial, traffic, performance, and operational data from more than 100 U.S. airlines.
- Initiate electronic filing: standardized electronic submission of data from the air carriers and the subsequent processing of the data in a centralized system that automates previous manual processes. If the USDOT expands current tarmac delay reporting requirements to all carriers, more resources for data collection and processing will be required.
- New IT system: convert existing legacy system into unified information system to enhance timeliness and data quality.
- Training: Instruct airlines on new data reporting requirements and processes.

Key Outputs:
- Faster, more accurate standardized data collection, eliminating manual data entry errors.
- Scheduled release of comprehensive on-line monthly, quarterly, and annual data on the operations of the airline industry.

Key Outcomes:
- Electronic data filing will provide USDOT officials, other decision-makers and stakeholders with more timely and accurate data for regulatory actions.
- Data form the basis for a greater understanding of the airline industry, which can shape decisions on safety, economic competitiveness, and improved operations.
- Electronic reporting will reduce reporting burden on airlines.

What Is This Program?

Strategic Goal: Safety, economic competitiveness, livable communities, environmental sustainability, state of good repair, and organizational excellence.

Purpose/Beneficiaries: The comprehensive program data necessary for consumer protection and enforcement activities, a major Secretarial priority, and for aviation policy decision making.

Partners: Stakeholders include:
- Decision-makers: The RITA Office of Airline Information (OAI) makes data available which provide policy-makers with robust, accurate data upon which to base decisions.
- USDOT Users:
    a) USDOT's Office of the General Counsel, Enforcement and Proceedings uses the data to monitor airline compliance with consumer protection regulations, and collaborates on data dissemination through the Air Travel Consumer Report and other vehicles; and
    b) The U.S. DOT Office of Aviation of International Affairs (OST-X) Office of Aviation Analysis uses the data to monitor and evaluate air carrier performance, economic fitness and competitiveness, and the Essential Air Service (EAS) program.
    c) The Federal Aviation Administration (FAA) uses the airline data to administer safety and airport improvement programs. The air carrier industry provides the data to BTS as required by regulation.
- Air travel consumer groups, media outlets and other stakeholders use BTS data to monitor and report on airline performance.
- State and local governments and regional airport authorities use BTS airline data to make strategic investment decisions.
- Airlines, consulting firms, and other entities in aviation use BTS airline data to help better align air transportation services with customer needs.

Description: The program collects and disseminates airline data related to on-time flights and other consumer-related issues, domestic and international passenger and freight traffic, passenger ticket information, and airline financial, fuel cost and consumption, and employment information. The Bureau of Transportation Statistics (BTS) collects data from 130 U.S. airlines and foreign carrier operations to and from the U.S. Annually, BTS collects more than 8,000 data reports from U.S. and foreign carriers. These data are reported by the air carriers as required by regulation.

Base Budget: The base budget is $4M; funding is provided by the FAA through an interagency agreement.

Prior Year Accomplishments:
- BTS implemented a data collection on tarmac delay and flight cancellations as required under new rulemaking to provide more detailed data on flight performance to air travel customers.
- Monthly, quarterly and annual airline financial, employment, and fuel cost and consumption data used to gauge the strength of the industry and individual airlines;
- Monthly air traffic reports on passenger enplanements, flights, freight and mail that signal how the airline system is operating;
- Monthly performance data that measures flight delays and cancellations, causes, bumpings and mishandled baggage reports; and
- Quarterly ticket data used to track airline fares and itineraries.

Why Is This Particular Program Necessary?

The program is the sole source of the airline operations data required by federal statute and regulation and essential for USDOT policy making, Congress, the airline industry and the traveling public. Without this funding:
- USDOT agencies and Congress would not have the data necessary to provide oversight and make informed policy decisions regarding the airline industry, including operational safety, and its impact on the economy and traveling public;
- The airline industry would lose the ability to schedule and set ticket prices based on objective, industry-wide airline data, with potential industry cost impacts since no private sector entity has the authority to collect industry-wide airline data; and
- The public would lose access to consumer information possibly used to make informed travel decisions such as airline on-time performance.

How Do You Know The Program Works?

Effectiveness Measures: The collection and reporting of airline on-time performance data, including its tarmac delay data, are key components of Secretary LaHood's successful efforts to ensure consumers have available the information necessary to make informed decisions about their travel, and to provide airlines an incentive to reduce the inconvenience and added expense to passengers that results from flight delays. The data also provide USDOT and stakeholders important information on the prevalence of the issue to better inform decision-making.

USDOT enforcement relies on data to enforce rules on tarmac times, chronically delayed and overbooked flights, and the Air Travel Consumer Report. Airline data are the most requested category on the BTS website with more than 10,000 requests per month. Industry representatives use the data for their own analyses and maintain links to the BTS webpage. BTS also receives information on customer satisfaction from the American Customer Satisfaction Index. BTS tracks air travel customers registered on social media networks such as Twitter. GAO relied heavily on BTS airline statistics for their study of airplane delays and study of ancillary revenues. (See GAO-10-542 *National Airspace System: Setting On-Time Performance Targets at Congested Airports Could Help Focus FAA Actions.*)

Research: Customer outreach; American Customer Satisfaction Index; data requests; and downloads from RITA web pages and stakeholder feedback.

Efficient: Federal statute and regulation require this data collection program. A key measure of efficiency for the air transportation statistics program is the timeliness with which BTS releases the air carrier data (passenger, flights, freight) to the public, typically within 30 days. On-time data are released within 15 days of carrier submission.

Why Do We Want/Need to fund The Program At the Requested Level?

Funding Options: To achieve a higher level of performance and usability from the airline data program; $5M is required for FY 2012. The increase of $1M over the FY 2010 funding level will support data modernization and related efforts, starting with an information systems requirements analysis that will define what will be needed in the future for development and implementation of the modernized system.

History/outyear needs: FY 2010 $4M; FY 2012 $5M.

THIS PAGE HAS BEEN INTENTIONALLY LEFT BLANK

EXHIBIT IV-1

RESEARCH AND INNOVATIVE TECHNOLOGY ADMINISTRATION
RD&T BUDGET AUTHORITY
($000)

Research and Development	FY 2010 Actual	FY 2012 Request	FY 2012 Applied	FY 2012 Development
Salaries and Administrative Expenses	4,701	4,385	3,288	1,097
Alternative Fuels Research & Development (R&D)	500	500	200	300
RD&T Coordination	536	900	657	243
Positioning, Navigation, and Timing	400	1,000	330	670
Intelligent Transportation Systems: Core Program [1]	[102,850]	[110,000]	[96,100]	-
University Transportation Center (UTC) Multimodal Competitive Research Grants [1]	-	[20,000]	[10,000]	[10,000]
Multimodal Innovative Research Program [1]	-	[20,000]	[10,000]	[10,000]
Total RITA	**6,137**	**6,785**	**4,475**	**2,310**

[1] Resources are shown as non-adds because the funding resides in the FHWA budget.